Intersectionality and Difference in Childhood and Youth

T0132529

This book explores the alternative experiences of children and young people whose everyday lives contradict ideas and ideals of normalcy from the local to the global context.

Presenting empirical research and conceptual interventions from a variety of international contexts, this book seeks to contribute to understandings of alterity, agency and everyday precarity. The young lives foregrounded in this volume include the experiences of transnational families, children in ethnic minority communities, street-living young people, disabled children, child soldiers, victims of abuse, politically active young people, working children and those engaging with alternative education. By exploring 'other' ways of being, doing, and thinking about childhood, this book addresses questions around what it is to be a child and what it is to be marginalised in society. The narratives explore the everydayness and the mundanity of difference as they are experienced through social structures and relationships, simultaneously recognizing and critiquing notions of agency and power.

This book, including a discussion resource for teaching or peer reading groups, will appeal to academics, students and researchers across subject disciplines including Human Geography, Children's Geography, Social Care and Childhood Studies.

Nadia von Benzon is a Lecturer in Human Geography at the University of Lancaster, UK.

Catherine Wilkinson is a Senior Lecturer in Education at Liverpool John Moores University, UK.

Routledge Spaces of Childhood and Youth Series
Edited by Peter Kraftl and John Horton

The *Routledge Spaces of Childhood and Youth Series* provides a forum for original, interdisciplinary and cutting edge research to explore the lives of children and young people across the social sciences and humanities. Reflecting contemporary interest in spatial processes and metaphors across several disciplines, titles within the series explore a range of ways in which concepts such as space, place, spatiality, geographical scale, movement/mobilities, networks and flows may be deployed in childhood and youth scholarship. This series provides a forum for new theoretical, empirical and methodological perspectives and ground-breaking research that reflects the wealth of research currently being undertaken. Proposals that are cross-disciplinary, comparative and/or use mixed or creative methods are particularly welcomed, as are proposals that offer critical perspectives on the role of spatial theory in understanding children and young people's lives. The series is aimed at upper-level undergraduates, research students and academics, appealing to geographers as well as the broader social sciences, arts and humanities.

Children, Nature and Cities
Rethinking the Connections
Claire Freeman and Yolanda van Heezik

Young People, Rights and Place
Erasure, Neoliberal Politics and Postchild Ethics
Stuart C. Aitken

The Common Worlds of Children and Animals
Relational Ethics for Entangled Lives
Affrica Taylor and Veronica Pacini-Ketchabaw

Intersectionality and Difference in Childhood and Youth
Global Perspectives
Edited by Nadia von Benzon and Catherine Wilkinson

For more information about this series, please visit: www.routledge.com/ Routledge-Spaces-of-Childhood-and-Youth-Series/book-series/RSCYS

Intersectionality and Difference in Childhood and Youth

Global Perspectives

Edited by
Nadia von Benzon and Catherine Wilkinson

Routledge
Taylor & Francis Group

LONDON AND NEW YORK

First published 2019 by Routledge

2 Park Square, Milton Park, Abingdon, Oxon, OX14 4RN
605 Third Avenue, New York, NY 10017

Routledge is an imprint of the Taylor & Francis Group, an informa business

First issued in paperback 2020

British Library Cataloguing-in-Publication Data
A catalogue record for this book is available from the British Library

Library of Congress Cataloging-in-Publication Data
A catalog record has been requested for this book

ISBN: 978-1-138-60829-0 (hbk)
ISBN: 978-0-367-72916-5 (pbk)

Typeset in Times New Roman
by Taylor & Francis Books

CW – To anyone who looks or feels different.
NvB – For my own children: Isabel, Anna Maria, Katie and Xavier; and for my god daughter Lucy.

Contents

Figures

Contributors

Jiniya Afroze, Open University

Nadia von Benzon, Lancaster University

Michael Boampong, Birkbeck, University of London

Erica Burman, University of Manchester

Sarah Marie Hall, University of Manchester

Urszula Markowska-Manista, University of Warsaw

Korinna McRobert, University of Applied Sciences Potsdam

Susie Miles, University of Manchester

Amy Mulvenna, University of Manchester

Gemma Pearson, Royal Holloway, University of London

Laura Pottinger, University of Manchester

Bernadine Satariano, University of Malta

Saskia Warren, University of Manchester

Catherine Wilkinson, Liverpool John Moores University

Samantha Wilkinson, Manchester Metropolitan University

Chris Willman, Open University

Helen F. Wilson, Durham University

Foreword

Intersectionality is, to date, a well-established theoretical and methodological tool to explore the experiences of childhood and youth. Yet, despite surges in intersectionality analyses within children's geographies, there has been, so far, limited attempts to bring together the multiple 'voices' and experiences of young people who live on the margins of society due to the history, culture and political economy of their own locales and beyond.

In this book, we see new ways of engaging with intersectionality in order to both excavate and 'bind' the lives of children and young people in diverse social, spatial, economic, institutional and political contexts. Implicit in the book is a critique of the neoliberal, agentic, unitary child subject as well as discourses of rights that romanticise children's competence and agency. We also learn a great deal about the hegemonic and 'globalizing' notions of childhood innocence, dependence, and domesticity that have neither existed in history nor characterise contemporary children's lives, but that yet inform policies and interventions that paradoxically pathologise 'disadvantaged' children and, in so doing, perpetuate their precarious existence.

The book's focus is not only on how children who are 'out-of-place' challenge expectations of childhood and whether their lives epitomise ambiguous childhood experiences worldwide. The book asks whether children's work, street existence, transnational migration, disability, sexuality, soldiering, being not in schools etc. calls forth an understanding of alterity rather than abnormality in childhood research. Contributions explore children's knack for and tenacity in overcoming the constraints of childhood; the tension between precarity and agency as well as how parameters of social differences such as gender, class, geography, disability, race and ethnicity speak to each other and, at times, intersect to operate as systems of exclusion, shaping what it means to be young in today's world. Contributions also explore the ways in which structural, discursive and political forces 'produce' precarious childhoods as well as manifestations of marginality linked to growing up in contexts of inequalities. They present and discuss complex adversities experienced by children; their agency in mustering resources for enduring poverty and, not least; how these issues are encountered and articulated by young people themselves. These explorations are not only long overdue but also deserve more attention than hitherto has been given to them in geographies of childhood and youth.

Of particular concern is the book's attentiveness to interpenetrating factors that affect children's multiple yet common experiences, revealing *processes* and *contexts* through which 'otherness' is generated; as well as savvy elements of affect, emotions, ethics, dilemmas and entanglements engendered by researching vulnerable children's everyday life. Children who are cast as 'abnormal' or simply 'outside childhood' will no longer be 'marginal' in debates around what it means to be young but, instead, recognised as experienced and 'real'. A unique contribution is the new spin the book makes upon the ideas, spaces and places such young people inhabit, not just as mere sites of marginalization and exclusion but, more importantly, as sites of resistance and activism through which they make and remake themselves.

For geographies of children and young people, therefore, this book represents an important leap forward. It reifies the paradox of difference implicated in social constructionism by revealing how young people's experiences of otherness 'jumps scale', attending to the spatialities of their localised, albeit interconnected realities, relationships and identities. The engagement of chapters that draw on reflexive, participatory and ethnographic research uncovers multiple exclusions borne by children and young people, and how 'the local' and 'the global' intertwine. Other chapters open space for a discussion around social and spatial justice for and with young people. The book demonstrates the potentials of analyses on the commonalities and differences in disadvantaged children's everyday lives, painting a more nuanced and complete picture of 'unconventional' childhoods and youth. The take away message for social policy is clear. By revealing some of the root causes of vulnerability and marginality, the book highlights arenas for collective action that goes beyond individuating children's adversity and resilience.

Tatek Abebe,
December 2018,
Trondheim, Norway.

Acknowledgements

Nadia and Catherine would like to thank each other. Catherine would like to thank Nadia for inviting her on board – Nadia would like to thank Catherine for agreeing to embark and for co-skippering the ship so enthusiastically and effectively. Most of all, it seems a good sign that they're still talking after 18 months of co-editing. Both editors should also thank The Tea Hive in Chorlton (www.teahive.co.uk) for providing a first rate venue and superb catering for editing meetings.

On a more serious note, we'd both also like to thank the many authors who generously contributed to this volume; we're really proud of the final edited collection. We have no doubt that the authors themselves owe a debt of gratitude to their participants, peers and mentors who have contributed to the development of the individual contributions. We know we certainly do.

Catherine would like to thank her parents, sister and partner for their continued support of her (ad)ventures both inside and outside of academia.

Nadia would like to thank/plug the immensely powerful network of support that is the Women in Academia Support Network. She'd also like to thank her husband for doing more than his share of the weekly food shop, and her kids who are really close to letting her get a full night's sleep.

1 Introduction

Nadia von Benzon and Catherine Wilkinson

This book is concerned with intersectionality and difference in childhood, presenting empirical research and conceptual interventions from a variety of international perspectives. Building on the now burgeoning body of discourse concerning intersectionality within feminist and broader social geography, this book seeks to explore the lived experiences of children facing 'multiple discriminations' (Valentine, 2007, p. 10) or stigmatised experiences in their day-to-day lives. Whilst the feminist concern at the heart of intersectionality has been gender,[1] the pivotal discriminatory identity at the heart of this volume is childhood. Children's geography, alongside allied social science disciplines, has sought in recent decades to demonstrate the ways in which the social constructions of childhood, alongside the material reality of a juvenile physiology and psychology, have led to the systematic and structural marginalisation of children and young people in societies (Valentine, 1996; Matthews, 2000; Collins and Kearns, 2001). The focus of this book is the lives of children and youth who face additional marginalisation, through their association with stigmatised identities or activities, or through their engagement with activities that are not widely deemed normalised or acceptable for children's participation.

There is a growing area of work in the study of childhood which has been concerned with childhoods that are 'different' or 'other' (see for example, the special edition of *Children's Geographies* in 2011 *Diverse Spaces of Childhood and Youth: Gender and socio-cultural difference* edited by Ruth Evans and Louise Holt; the 2017 special edition of the same journal *Intersectionality in Childhood Studies* edited by Kristina Konstantoni and colleagues; Kraftl, 2015). Whilst alterity does not necessarily lead to stigma and marginalisation (your childhood could be 'other' due to participation in elite activities, access to exclusive spaces or possession of prodigious abilities), difference for any reason may render a child vulnerable. This vulnerability is due to their potential lack of inclusion in policies and practices designed to safeguard children framed within a notional construction of a 'normal' childhood, or the fragility and alterity of their relationships with other children and adults. The vulnerability inherent in difference in childhood has been painfully illustrated in the #MeToo movement which has seen (former) child actors and athletes coming forward to expose the perpetrators who were able to assault them due to the power they held over children who were trying to succeed in an adult-dominated environment, or who felt intimately reliant on them for their coaching and ultimate success.

However, research addressing childhood difference, particularly from geographers, has typically focused on differences that may manifest in stigma and marginalisation. The exception to this in terms of a conflation of difference and marginalised experience has largely concerned education, particularly with Peter Kraftl's recent work on alternative education (Kraftl, 2014; Kraftl, 2015), and some reflection on youth engagement with politics (e.g. Skelton, 2010). The primary geographical focus of difference in terms of stigma and marginalisation is likely to reflect a political attempt by geographers to listen to children with the quietest voices, and a desire to magnify these children's voices for emancipatory ends. Thus, an imperative towards social justice, alongside the development of social theory – or academic pontification – underpins much of the children's geographies research addressing lived experiences of difference (Holt, 2007; Beazley et al., 2009; Klocker, 2011; Bradbury-Jones and Taylor, 2015).

Key themes within this broad children's geographies literature on marginalised childhoods have included: race, ethnicity and religion (Dwyer, 1999; Hopkins et al., 2017); disability, health, and caring responsibilities (Robson, 2000; Holt, 2010; Worth, 2013; Pyer and Tucker, 2017; von Benzon, 2017a); gender and sexuality (Skelton, 2000; de Montigny and Podmore, 2014); and poverty and class (McDowell, 2003; Pimlott-Wilson and Hall, 2017). The work of Maria Rodó-de-Zárate (2017) demonstrates that for some children and young people, it is their non-age-related identity, particularly their sexuality, ethnicity and gender, that provokes considerably more discrimination than their youthful age. Indeed, Hopkins and Pain (2007, p. 290) argue: 'some markers may intersect with age in very powerful ways; others may make age far less significant in relation'. Thus, geographers have demonstrated the potential of a hierarchy of identities, but stress (Valentine, 2007) that the articulations between social categories are complex and dynamic and contingent upon time and space. Many of these themes are covered in the chapters herein as they seek to address alternative experiences of childhood that contest the construction of a 'normal' childhood which dominates within a particular community or locality.

The chapters in this volume might, to some extent, be considered accidental interventions in the development of intersectionality theory. Unlike Valentine (2007) and Rodó-de-Zárate (2014), none of the chapters herein set out explicitly to explore the impact of intersectionality on individuals' lives nor do they knowingly balance or make judgements about the relative importance of different sorts of marginalisation, whether 'spoiled identity' or less detrimental lived experiences of difference. However, the chapters all, if only unwittingly, serve to respond to Hopkins' (2018) call to recognise the relationality and the chronological and spatial context of marginalisation. These chapters speak to the importance of locality in the way in which identity manifests as lived social experience. Bernadine Satariano's contribution to this volume is particularly powerful in demonstrating the way in which a particular identity – as a child of a 'broken' home, may manifest as 'spoiled' in one community, whilst in the same country, a matter of miles away, the same identity is considered normal, unremarkable, or even as positive. Thus, this volume addresses the issue of socio-spatial context as a syphon for intersecting identity and experiences of difference, not only at an international scale but at a local and community scale. At once, culture is positioned as a crucial mediator in social experience of difference, whilst these lived experiences of difference are positioned as a lens through which localised cultures can be understood.

As a geographical contribution to this literature, *Intersectionality and Difference in Childhood and Youth* offers a critical exploration of research with children and young people who may be considered 'out of place' within their local context. Thus, this book approaches 'normalcy' and 'difference' as relational terms with the notion of the 'normal' child encompassing specific assumptions about agency, rights and responsibilities within different communities. Looking outside the global north, what constitutes a 'normal', 'mainstream' or 'acceptable' childhood may be a broader spectrum of experience, requiring a more nuanced ethical, methodological and theoretical interpretation. Whilst 'global expectations' might be a difficult concept to define or morally justify, one example illustrating such are the rights enshrined within the UN Convention on the Rights of the Child, the most widely ratified of all international conventions. Considering 'other' childhoods, both in the global north and the global south, can call into question many of the pre-conceived and socially constructed values and assertions concerning childhood that underpin national and institutional policies and dictate the ways in which children and adults interact. We hope that a broader recognition of the range of experiences of childhood, and a better understanding of the lived experience of these 'other' childhoods, will create space for rigorous debate around core geographical issues such as mobilities, agency, risk, work, education and play.

Much of the published work in this field to date has reflected on the difficulty to engage marginalised children in research – whether the difficulty of access through gatekeepers, social and spatial challenges in communicating directly and independently with children, and overcoming unbalanced power relations to develop researcher-child relationships with integrity (Barker and Weller, 2003; Lomax et al., 2011; Skovdal and Abebe, 2012; von Benzon, 2017b). These challenges are reflected on by authors in this volume, who discuss the assumptions made by participants that the research will bring a family material benefit (Boampong, Afroze), or the difficulty of accessing children's own internet-published material (von Benzon). Some chapters never made it to the final volume, due to irreducible challenges in undertaking the fieldwork, whilst others were almost curtailed due to issues over rights and permission. Thus, the empirical research and the conceptual interventions presented through this book are valuable – these narratives were hard won through engagement with fieldwork, or ideas, that were often deemed risky. In engaging with these children – through real-world interaction, documentary analysis and fiction, researchers themselves experience vulnerability; emotionally, bodily and cognitively. The risk, the precarity, and the emotional investment of research with marginalised children must become embedded within the pages of the subsequent publications. Whilst this investment will not, of course, have an automatic impact on the quality of the outputs, the cost of producing them must have an inherent impact on their ultimate value.

The book contributes to both the developing discussions concerning the precarious nature of marginalised children's lives, and the ways in which this precarity might be exacerbated or managed through the child's broader relationships with both people and non-human agents. We recognise children as independent actors capable of managing themselves in uncertain spaces, whilst also being influenced and constrained by broader social, political and economic factors over which they have no control. The chapters in this book are concerned with exploring the everydayness of precarity, revealing the reality of the emotional and embodied experience of precarity whilst also demonstrating how 'other' children routinely develop strategies that limit the experience of uncertainty in particular spaces. The authors in this volume do not shy away from the challenge of approaching some of the most ethically complex experiences of difference, uncertainty, agency and vulnerability faced by children in contemporary society. Chapters address child soldiers (Markowska-Manista), sexual assault (McRobert), child labour (Willman) and the lives of street-connected youth (Pearson).

In so doing, the volume seeks to emphasise the importance of focusing resources in support of those children whose lives differ from the normative, or at least, normalised, view of childhood present in the internationalised discourse that underpins child-focused regulation, policies and services. However, in opening discourse on everyday alterity and contingent precarity, the volume seeks to contribute to the growing geographical discourse calling for a queering of children's lives (Taylor and Blaise, 2014; Shillington and Murnaghan, 2016). That is a picking apart of the assumptions that are made, not just about children's sexuality, but about ability, opportunity and desire, each underpinned by children's intersectional identities of class, race, impairment, wealth, religion, and the many other axes of difference illuminated in the chapters of this book. The existence of so much difference – the vast heterogeneity of children's experiences, not only internationally but also at a highly localised scale, calls into question the utility of the notion of childhood itself – and certainly of the value of international and national policies that seek to bind particular lived experiences to the construct of 'the child'.

Through presentation of a variety of experiences of childhood that contradict the multiple and context-specific constructions of 'normal childhoods', *Intersectionality and Difference* seeks to demonstrate the way in which precarity and agency can be concurrently experienced in the lives of out-of-place children. The lives presented in this collection should neither be considered tragic, nor heroic, but everyday lives in which children are agential actors employing what capital they can access, and drawing on complex networks of human and non-human relationships, in order to address the challenges they face. As such many of these children are at once fragile, vulnerable to (further) upheaval, abuse or stigmatisation and also coping. Their capacity to 'get by' is based on a complex interweaving of personal resiliency, interpersonal support and sometimes structural interventions available through outside agencies. Ultimately though, agency suggests choice, and for many of the protagonists of this volume, choice, and thus meaningful agency, is limited (see Barker et al., 2010). This book presents a timely intervention in a period when the idea of agency, a core concept in the development of children's geographies to date, is being interrogated (Kraftl, 2013) and reframed (Holloway et al., 2018).

The book collates a series of different narratives exploring the ways in which experiences of childhood can counter the hegemonic narratives of normal childhood that are plural on a global-scale, but often singular at the local. This book moves away from a vision of a single, idealised, normalised childhood in favour of multiple childhoods, and considers the way in which children and young people experience childhood and youth in a variety of contexts. Looking from both majority and minority world perspectives, the book demonstrates children as both passive and active agents of their lived experience, illustrating the way in which agency is underpinned by social and economic capital and relationships, which vary according to geo-social contexts.

Structure of the book

Each contribution to this edited collection presents a discourse based on research addressing children who are marginalised and stigmatised due to difference, or for whom difference is the result of personal or familial choices that contradict social norms. The chapters each demonstrate the ways in which children's lives are shaped by intersecting identities and notions of difference. For these out of place children, their lived experience is as much about vulnerability and precarity as it is a story of tenacity, agency and the deployment of social capital. This book includes chapters that discuss the experiences of children and childhoods that are 'other'. Each individual chapter provides a lens on a particular context in which young people are constructed, sometimes by themselves, their families or through choice, but more often by the actions and discourse of wider society, as 'different'. The chapters focus in on a particular childhood identity or a particular event or activity experienced by a child that may mark them out as different. The chapters draw on a variety of geographical contexts, from Bolivian working and lower middle-class families, to minority ethnic children living in encampments in Bangladesh, to wealthy children in suburban Australia. The chapters also draw out a variety of key themes in children's geographies, including education, play, work and mobilities, both of relevance to those studying children and families, and of broader relevance to debates in human geography and across the social sciences.

Intersectionality and Difference is divided into three sections: I. Stigma; II. Work, Education and Activism; and III. Out of Place.

In Part I the authors draw on conceptualisations of stigma. They explicitly or implicitly examine the ways in which social perception of their identities as 'spoiled' renders children and young people different within local, national or global contexts. It is through this negative perception of difference that the young people in these chapters become marginalised. These chapters explore disability, street-connectedness, family structure and violence as intersections at which young people can experience precarity within their local contexts. The chapters demonstrate struggles between children's limited agency to improve their lived experience, and social attitudes that diminish their ability to lead the lives and experience the childhoods to which they aspire.

In Chapter 2 Catherine Wilkinson and Samantha Wilkinson consider the relationship between childhood disability and clothing. This chapter reviews literature to discuss the social and symbolic status of clothing – both mainstream clothing ('fashion') and adaptive clothing – for children and young people with a range of disabilities. In doing so, it provides an enhanced understanding of the lived experiences of these 'other' childhoods. This chapter argues that rehabilitation professionals need to take clothing-related issues faced by disabled children more seriously, so that opportunities for social participation can be maximised.

In Chapter 3 Gemma Pearson focuses on the lives of street-living children in Tanzania. This chapter argues that street-living children find themselves excluded from mainstream society due to the stigmatisation they receive for appearing to live independently of their parents and wider family in a societal context which prioritises intergenerational reciprocity and parental obedience. This chapter draws on the concept of intergenerational reciprocity, street child literature and empirical research to explore who is responsible for street-living children, how they are responsible and to what extent.

Bernadine Satariano, in Chapter 4, explores wellbeing differences based on socio-cultural differences in three Maltese neighbourhood communities. This chapter explores the way in which parental marital status is perceived in the neighbourhood's cultural and social context and how the child experiences the repercussions of familial status. It emphasises how much the history of the place shapes the norms which are socially accepted in the locality, and thus highlights that the health and wellbeing of the people is contingent to the place.

The focus of Chapter 5 is child soldiers. Recognising that training underage recruits to kill has been the central axis of many cases of military mobilisation, Urszula Markowska-Manista discusses the practices of training children to fight as soldiers and their aides in wars and armed conflicts on the African continent, with a particular focus on the previous decade. Markowska-Manista argues that, due to the psychosocially and politically negative nature of this type of influence, as well as the inhumane methods and unethical tools used in this educational process, this practice can be defined as a deliberate transgression of educational boundaries.

In Part II, authors look at intersectionality and difference in children in settings related to work, education and activism. The children in this section of the book are all political actors in one way or another, whether through unwitting participation in social movements, as targets of national policy or as unionised or organised agents in political processes. However, whilst this section of the book may highlight demonstrations of agency, it continues to focus on the intersections affecting these young politicised agents: particularly race and ethnicity.

In Chapter 6 Helen Wilson, Erica Burman, Susie Miles and Saskia Warren draw on four years of ethnographic research with supplementary schools in Manchester, UK. By situating the case of supplementary education within wider debates on the construction of educational 'problems' and 'multicultural schemas of perception', the chapter examines how forms of supplementary education have repeatedly been subject to scrutiny, surveillance, and multiple practices of othering that work to exclude 'other' childhoods, migrant histories, and forms of community.

The focus of Chapter 7 is children that work in Bolivia. Chris Willman draws on fieldwork to explore childhood in relation to child labour in the socio-political-legal context of Bolivia. It is a contribution to a new body of literature that looks at working children as opposed to child labour in its worst forms. The chapter argues that a working childhood in Bolivia is 'normal' within the national context, and would only be considered 'other' if approached from a global north perspective. It finds that Bolivia's new child labour law at once complements and rejects current international standards.

Chapter 8 draws on research addressing publicly available blogs published by mothers who home educate their children using the 'unschooling' approach. Nadia von Benzon argues that children who are unschooled can be considered 'other' within a neoliberal context in which formal education is seen as a building block for a healthy economy. This chapter considers the ways in which 'otherness' is recognised in unschooling families, and the techniques that are employed by these families to ensure that this arguably self-imposed 'otherness' is a positive experience for their unschooled children.

In Chapter 9, Sarah Hall and Laura Pottinger explore how and why young people's views, experiences, ideas and priorities matter in political processes and decisions. Drawing on ethnographic and participatory research the chapter considers how political issues and agendas are developed and led by young people and those who facilitate their engagement. Hall and Pottinger reveal how young people – particularly from working class backgrounds and black and minority ethnic communities – are being seen and being heard, notwithstanding the challenges they face to be taken seriously as political actors and activists.

The chapters in Part III consider the idea of being 'out of place'. This final section of the book highlights 'out of place'-ness or difference as both a social and a spatial occurrence – exploring the ways in which the social and the spatial intersect in a way that makes particular children not fit within the broader community in which they live, or within policies and processes.

In Chapter 10, Korinna McRobert looks at children 'out of place' as a category, focusing on the example of the sexually abused and thus sexually exposed child, as experiencing an 'out of place' childhood. The issue is contextualised by looking at the origins of the term 'children out of place' and how the sexually abused and exploited child can be framed and categorised in this way. The nature of the/a child-adult sexual relationship dynamic is analysed in terms of one case represented in the film *UNA* (2017).

Chapter 11 draws on ethnographic fieldwork conducted in an Urdu-speaking Bihari community in Dhaka, Bangladesh, to explore the lived experiences of children in a socio-spatial context where, due to their ethnic-linguistic background, which is compounded by poverty, children experience inequality and deprivation as part of their everyday lives. Jiniya Afroze frames childhood from a generational perspective, and presents data from both children and adults through semi-structured individual interviews, group discussions and participant observations. The findings of this research illustrate the ways adults perceive socio-spatial risks for children and the way children experience and construct private and public spaces.

In Chapter 12, Amy Mulvenna considers a tale by Australian author-illustrator Shaun Tan that deals imaginatively with an 'othered' young subject who negotiates and plays with place, space and *things* within an unfamiliar suburban environment. In Mulvenna's analysis of the text, she explores how Tan brings mapping, touch and materiality together metaphorically, imaginatively, and most intimately, through the character of Eric. The chapter examines ways in which Eric's intra-actions *with* and plotting *of* emergent im/materialities provide starting points for deviations from known routes.

Chapter 13 draws on fieldwork with Ghanaian transnational households to engage with the discussion on transnational migration processes, the practice of doing family, and children's role as social actors in periods of economic constraints. The focus is on the increasing number of children at different stages of family migration processes who support reproductive processes that help to mitigate the impact of economic constraints. Using data obtained from fieldwork on the everyday lives of transnational families, Michael Boampong analyses how children's presence and participation in imagined or physical migration processes, and how daily life practices of the transnational household, serve to constitute the transnational social field.

By way of a conclusion, Chapter 14 draws out the key conceptual themes in the book, focusing particularly on the way in which the empirical contributions might influence developments in framing the relational nature of agency. This chapter reflects on the utility of a scholarly discourse concerning intersectionality and difference in childhood, and childhood as a contested and precarious space.

The final section of the book, the Appendix, presents a resource designed for facilitating the use of the book in peer discussion groups and as a teaching tool. These questions are organised first relating to each chapter individually, second relating to the three parts of the book, and finally relating to the book as a whole. In this way the questions might be used with groups addressing the book at different scales, or might be useful in settings in which students or other discussants have read different chapters. Whilst devised for oral or online discussion groups or forums, some of the questions might also be useful as a way of guiding summative or formative written exercises or assessments. It is our hope that the Appendix provides a useful tool for engagement with the book in peer or student-teacher educational practice. Individuals using the book for their own interest or development may also be interested in reflecting on these questions in order to create time and space to engage with the key themes of the volume.

To summarise, this book examines intersectionality and difference in childhood. In doing so, it draws on local and global examples across a wide range of settings to detail the experiences of marginalisation and stigmatisation relating to children's perceived alterity. Many of the chapters in the book illustrate the way that alterity might be associated with vulnerability and precarity, sometimes related to 'personal' factors such as a child's body, family situation or other aspects of their identity, and often related to broader socio-political structures. The book therefore demonstrates a need for understanding of everyday precarity in children's lives, including the broad assemblages of relationships, experiences, social capital and structural constraints, in order to gain meaningful insight. The chapters in *Intersectionality and Difference* seek to paint a nuanced picture of the way in which children and young people may respond to marginalisation and stigma, demonstrating resilience and agency, and drawing on social and educational capital to affect change.

Note

1 Specifically the discrimination inherent in womanhood which is then magnified through dual or multiple memberships of other stigmatised, devalued or 'spoiled' identities (Goffman, 1973).

References

Barker, J., Alldred, P., Watts, M., & Dodman, H. (2010). Pupils or prisoners? Institutional geographies and internal exclusion in UK secondary schools. *Area*, 42(3), 378–386.

Barker, J., & Weller, S. (2003). 'Is it fun?' Developing children centred research methods. *International Journal of Sociology and Social Policy*, 23(1/2), 33–58.

Beazley, H., Bessell, S., Ennew, J., & Waterson, R. (2009). The right to be properly researched: Research with children in a messy, real world. *Children's Geographies*, 7(4), 365–378.

Bradbury-Jones, C., & Taylor, J. (2015). Engaging with children as co-researchers: Challenges, counter-challenges and solutions. *International Journal of Social Research Methodology*, 18(2), 161–173.

Collins, D. C., & Kearns, R. A. (2001). Under curfew and under siege? Legal geographies of young people. *Geoforum*, 32(3), 389–403.

de Montigny, J., & Podmore, J. (2014). Space for queer and trans youth? Reflections on community-based research in Montreal. *Global Studies of Childhood*. doi:10.2304/gsch.2014.4.4.198

Dwyer, C. (1999). Contradictions of community: Questions of identity for young British Muslim Women. *Environment and Planning A*, 31, 53–68.

Evans, R. and Holt, L. (2011). Diverse spaces of childhood and youth: Gender and other socio-cultural differences. *Children's Geographies*, 9(3–4), 277–284.

Goffman, E. (1973). *The Presentation of Self in Everyday Life*, Woodstock, NY: Overlook Press.

Holloway, S., Holt, L., & Mills, S. (2018). Questions of agency: Capacity, subjectivity, spatiality and temporality. *Progress in Human Geography*. doi: doi:10.1177/0309132518757654

Holt, L. (2007). Children's sociospatial (re)production of disability within primary school playgrounds. *Environment and Planning D: Society and Space*, 25(5), 783–802.

Holt, L. (2010). Young people's embodied social capital and performing disability. *Children's Geographies*, 8(1), 25–37.

Hopkins, P. (2018). Feminist geographies and intersectionality. *Gender, Place and Culture*, 25(4), 585–590.

Hopkins, P., Botterill, K., Sanghera, G., & Arshad, R. (2017). Encountering misrecognition: Being mistaken for being Muslim. *Annals of the Association of American Geographers*, 107(4), 934–948.

Hopkins, P., & Pain, R. (2007). Geographies of age: Thinking relationally, *Area*, 39(3), 287–294.

Klocker, N. (2011). Negotiating change: Working with children and their employers to transform child domestic work in Iringa, Tanzania. *Children's Geographies*, 9(2), 205–220.

Konstantoni, K., Kustatscher, M., & Emejulu, A. (2017). Travelling with intersectionality across time, place and space. *Children's Geographies*, 15(1), 1–5.

Kraftl, P. (2013). Beyond 'voice', beyond 'agency', beyond 'politics'? Hybrid childhoods and some critical reflections on children's emotional geographies. *Emotion, Space and Society, 9,* 13–23.

Kraftl, P. (2014). *Geographies of Alternative Education: Diverse learning spaces for children and young people*, Bristol: Policy Press.

Kraftl, P. (2015). Alter-childhoods: Biopolitics and childhoods in alternative education spaces. *Annals of the Association of American Geographers*, 105(1), 219–237.

Lomax, H., Fink, J., Singh, N., & High, C. (2011). The politics of performance: Methodological challenges of researching children's experiences of childhood through the lens of participatory video. *International Journal of Social Research Methodology*, 14(3), 231–243.

Matthews, H., Taylor, M., Percy-Smith, B., & Limb, M. (2000). The unacceptable flaneur: The shopping mall as a teenage hangout. *Childhood*, 7(3), 279–294.

McDowell, L., (2003). *Redundant masculinities: Employment change and white working class youth*, Oxford: Blackwell.

Pimlott-Wilson, H., & Hall, S. M. (2017). Everyday experiences of economic change: Repositioning geographies of children, youth and families. *Area*, 49(3), 258–265.

Pyer, M. and Tucker, F. (2017). 'With us, we, like, physically can't': Transport, mobility and the leisure experiences of teenage wheelchair users. *Mobilities*, 12(1), 36–52.

Robson, E. (2000). Invisible carers: Young people in Zimbabwe's home-based health-care. *Area*, 32(1), 59–69.

Rodó-de-Zárate, M. (2014). Developing geographies of intersectionality with Relief Maps: Reflections from youth research in Manresa, Catalonia. *Gender, Place and Culture*, 21(8), 925–944.

Rodó-de-Zárate, M. (2017). Who else are they? Conceptualizing intersectionality for childhood and youth research. *Children's Geographies*, 15(1), 23–35.

Shillington, L. J., & Murnaghan, A. M. F. (2016). Urban political ecologies and children's geographies: Queering urban ecologies of childhood. *International Journal of Urban and Regional Research*, 40(5), 1017–1035.

Skelton, T. (2000). 'Nothing to do, nowhere to go?': Teenage girls and 'public' space in the Rhondda Valleys, South Wales. In *Children's Geographies: Playing, living, learning*, S. Holloway & G. Valentine (eds.). 80–99. New York and London: Routledge.

Skelton, T. (2010). Taking young people as political actors seriously: Opening the borders of political geography. *Area*, 42(2), 145–151.

Skovdal, M., & Abebe, T. (2012). Reflexivity and dialogue: methodological and socio-ethical dilemmas in research with HIV-affected children in East Africa. *Ethics, Policy & Environment*, 15(1), 77–96.

Taylor, A., & Blaise, M. (2014). Queer worlding childhood. *Discourse: Studies in the Cultural Politics of Education*, 35(3), 377–392.

Valentine, G. (1996). Angels and devils: Moral landscapes of childhood. *Environment and Planning D: Society and Space*, 14(5), 581–599.

Valentine, G. (2007). Theorizing and researching intersectionality: A challenge for feminist geography. *The Professional Geographer*, 59(1), 10–21.

von Benzon, N. (2017a). Unruly children in unbounded spaces: School-based nature experiences for urban learning disabled young people in Greater Manchester, UK. *Journal of Rural Studies*, 51, 240–250.

von Benzon, N. (2017b). Confessions of an inadequate researcher: Space and supervision in research with learning disabled children. *Social & Cultural Geography*, 18(7), 1039–1058.

Worth, N. (2013). Making friends and fitting in: A social-relational understanding of disability at school. *Social & Cultural Geography*, 14(1), 103–123.

Part I
Stigma

2 Childhood disability and clothing

(Un)dressing debates

Catherine Wilkinson and Samantha Wilkinson

Chapter summary

This chapter considers the relationship between childhood disability and clothing. It stems from the understanding that a physical disability may be used as a cue to categorise a person as abnormal, different, or indeed 'other', and considers how clothing is used in the negotiation and presentation of self. This chapter reviews literature to discuss the social and symbolic status of clothing – both mainstream clothing ('fashion') and adaptive clothing – for disabled children. In doing so, it provides an enhanced understanding of the lived experiences of these 'other' childhoods. A variety of normalising techniques are discussed throughout this chapter. For instance, 'making do' with ready-to-wear clothing through resourceful adaptations; deflecting attention from a disability; and concealment of the disability using clothing. Collating scholarship from a range of disciplines (including but not limited to geography, sociology, disability studies, health, and clothing and textiles studies) this chapter argues that rehabilitation professionals need to take clothing-related issues faced by disabled children seriously (as opposed to seeing them as solely aesthetic), so that opportunities for social participation can be maximised.

Introduction

Whilst geographers' interest in the subject of disability has traditionally been marginal (Worth, 2008), in more recent decades geographical scholarship about space, place and disability has proliferated (Imrie and Edwards, 2007). Within the subdiscipline of children's geographies, scholars have shown interest in the experiences of disabled children and young people (e.g. Hodge and Runswick-Cole, 2013; Holt, 2004; 2006; 2010; Pyer et al., 2010; Valentine and Skelton, 2003; 2007; von Benzon, 2010; 2017a; 2017b). By and large, this research has found disabled children and young people to be excluded, both socially and spatially, as a result of stigmatisation through perceived difference, and as a failure to be able to perform their identity in the same way as their peers (von Benzon, 2017b).

Childhood disabilities are conditions that are highly likely to affect the trajectories of children's development into adulthood. Many have a neurological basis, whilst others may include musculoskeletal conditions or genetic syndromes, and cognitive, behavioural and communication disorders (Rosenbaum and Gorter, 2011). There are two widely recognised approaches to disability, the 'medical' or 'individual tragedy' model and 'the social model' (Parr and Butler, 1999). Whilst it is not the purpose of this chapter to delineate each of the models (we refer readers to Goodley and Runswick-Cole, 2010 for a comprehensive discussion of theorising disabled childhoods) we will briefly outline these two models here. The medical model of disability has a focus on curing the 'deviant' body or mind (Oliver, 1996), believing that problems arise from deficits in the body (Shakespeare and Watson, 2001). The social model sees society's attitudes as the main problem for disabled people (as opposed to the disability itself). This model describes the social barriers which prevent equal opportunity, believing these can be overcome by social change (Saunders, 2000). This chapter adopts the stance of the social model of disability. We acknowledge that the social model is not without critique; for instance, social model theorists have paid limited attention to disabled children, with a focus on adults, and arguably under-represent 'difference' between disabled people (Connors and Stalker, 2007). However, we adopt this model as it destabilises the location of the cause of disability from the individual to society.

Textiles and clothing scholars have addressed concerns of disabled wearers since the 1950s, although, as Lamb (2001) points out, early research on clothing for disabled wearers was dominated by a medicalised view of disability. While some writing (e.g. Ratna Jyothi, 1988; Singh and Ghai, 2009) has argued that garments should disguise the disability to the extent possible, a social model of disability is compatible with textiles and clothing scholarship on appearance and social realities (Lamb, 2001) and would argue that clothing designers and manufacturers should accommodate the needs of disabled people to enable them to have full participation in society. For disabled wearers, clothing should be easy to put on and take off (facilitating independence in dressing), and comfortable and non-restricting (allowing movement and mobility). The social view of disability suggests that clothing designers would be more successful if they involved disabled consumers throughout the design process. Clothing plays a significant role in contemporary life and understanding this role in the lives of disabled children is important. This importance is particularly apparent when considering Kabel et al.'s (2016) powerful argument that lack of appropriate clothing can stop disabled people from fully engaging in everyday life, to the same extent as unsuitable pavement curbs and doorways.

This chapter discusses the social and symbolic status of clothing – both mainstream clothing ('fashion') and adaptive clothing – for disabled children and young people. It is concerned with exploring the reality of the embodied experiences of everyday life for these children and young people, whilst also demonstrating how such 'other' children develop strategies that limit the experience of uncertainty and 'deal with' looking and feeling different. A range of physical disabilities are referred to in this chapter, including: Down's Syndrome; spina bifida;[1] osteogenesis imperfecta;[2] and visual impairments. Kabel et al. (2016, p. 2184) have argued that the relationship between clothing and marginalisation of disabled people is 'powerful yet often invisible'. Through making this relationship visible, this chapter makes important contributions to understandings of the alternative childhoods of disabled children.

This chapter proceeds as follows. First, we present an overview of literature on children and clothing. We then contextualise the discussion by outlining debates on conceptualisations of childhood disability. Following this, we draw together studies on childhood disability, dress, and the body. In doing so, we consider the extent to which disabled children are othered by dress or, alternatively, use dress to feel less othered. We break this section down into four thematic areas: finding appropriate clothing; clothing and the management of identity; disability, clothing and mobility; and dress and undressing. The chapter concludes with a call for rehabilitation professionals to take clothing-related issues faced by disabled children seriously – as opposed to considering them purely aesthetic – so that opportunities for social participation can be maximised.

Children and clothing

Most children and young people have independent entry into social and cultural life, for instance through consumerism and fashion, thus offering opportunities for them to 'do' their identity differently (Valentine, 2000). In producing their own narrative of self, children learn to negotiate their identity to position themselves correctly within adult and peer cultures (Valentine, 2000). We already know that making decisions about how to dress draws on personal creativity, but also on social constraint, and that clothing's semiotic and sensual material propensities embody conventions about propriety, gender, ways of moving, and encode social relationships, status, biographies and identities (Candy and Goodacre, 2007). In this sense, children's fashioned bodies act as a site through which they explore and express their self-identity.

Research into fashion and children has covered broad ground; for instance the commodification of children (Cook, 2004); the production of the 'profitable child' (Crewe and Collins, 2006, p. 7); branding and children (Ross and Harradine, 2004); symbolic consumption in clothing choices (Piacentini and Mailer, 2004); and the intergenerational gap of interpretation of young girls' clothes (Rysst, 2010), including 'sexy girls' clothes' (Torrell, 2004). Piacentini and Mailer (2004) find that the clothing choices made by young people are closely bound to their self-concept,[3] and are used both as a means of self-expression and as a way of judging the people and situations they face. Findings in this research also suggest that clothing has a function in role fulfilment, making the wearer more confident and capable of performing tasks. In sum, clothing can be viewed as an essential social tool in the lives of children and young people. This considered, the role of clothing in the lives of disabled children (who often worry about social situations and experience social exclusion) is an important area of attention.

A major body of work has considered identity in relation to children and clothing. For instance, Boden (2006) discusses how popular culture can influence children's social identities, self-styling and the presentation of their identity. Pilcher (2010) has discussed young girls, clothing and 'showing' the body. Drawing on an ethnographic study of children aged 6–11 years and their families, Pilcher (2010) presents girls' constructions of fashion in relation to their own bodies and those of others. It is shown that although girls may desire to 'dress up' in fashionable clothing, they present contradictory meanings for doing so. For some girls, 'dressing up' in certain clothes may be a way of 'ageing up' toward feminine adulthood, albeit in restricted contexts and after negotiations between themselves and their parents as to what can be worn and where. Young girls in the study also showed anxieties and disapproval of 'showing the body' through revealing clothing. Certainly, anxieties around showing the body, or particular aspects thereof, through revealing clothing may also be apparent for some disabled children.

Childhood disability

In recent decades, there have been significant developments in research, policy and practice relevant to the lives of disabled children (Kelly and Byrne, 2015). At an international level, disabled children have been recognised as rights holders under both the United Nations Conventions on the Rights of the Child (1989) and the United Nations Convention on the Rights of Persons with Disabilities (2006). These international frameworks highlight the intersections between childhood and disability and affirm the rights of disabled children to protection, participation, and provision of relevant services and support. Despite this, disabled children and young people continue to experience disadvantage, including higher levels of poverty, lower educational attainment, limited engagement in employment, poorer health outcomes (Kelly and Byrne, 2015) and social exclusion (Lindsay and McPherson, 2012).

Discussing 'notions of self' and the lived realities of disabled children in India, Singh and Ghai (2009) find that disabled children desired to appear similar to 'non-disabled' children. Within the Indian context, disabled children have typically been categorised as silent, voiceless victims (Corker and Davis, 2000). This led Singh and Ghai (2009) to argue that disabled children must be understood as social actors, as controllers, and as negotiating their complex identities within a disabling environment. In order to find a place for disabled children in a social and cultural context that has historically cast them as 'monstrous others', Goodley et al. (2016, p. 771) develop the theoretical notion of the 'DisHuman'; a bifurcated being that allows us to recognise their humanity whilst also celebrating the ways in which disabled children reframe what it is to be human. The authors suggest that the lives of disabled children and young people demand us to think in ways that affirm the inherent humanness in their lives but also allow us to consider their disruptive potential. As can be seen, the theorising of childhood disabilities is a contested terrain.

Few studies have focussed specifically on children's perceptions and experiences of disability, with much research speaking to parents, and principally mothers, to understand their experiences of looking after their child. For instance, the constructing of daily routines between mothers and disabled children (e.g. Kellegrew, 2000), and of parents' facilitation of friendships between their disabled child and friends without a disability (Turnbull et al., 1999). Arguably, the difficulty to gain ethical approval to speak to disabled children directly can be a reason for this lack of direct engagement. However, there are exceptions (for a recent exception see Runswick-Cole et al., 2018) and some researchers have adopted different methodologies and methods when researching about/with disabled children to deepen their understandings. Davis, Watson and Cunningham-Burley (2008) adopt an ethnographic approach to 'get at' the 'unspoken understandings' of disabled children. Wickenden and Kembhavi-Tam (2014) use participatory research to actively include disabled children and young people in research that explores their lives. Pickering (2018) uses diaries and interviews with disabled children and young people to gauge their opinions on adapted cycling. These efforts to include disabled children in research are important as we have much to learn about their own experiences, including in the domain of clothing. In the discussion that follows, we focus on four thematic areas in the literature: finding appropriate clothing; clothing and the management of identity; disability, clothing and mobility; and dress and undressing.

Finding appropriate clothing

Caregivers of disabled people have long reported the difficulties in finding clothing with desired features, such as elastic waistbands that are age, size and situation appropriate (Watson et al., 2010). Writing more than 20 years ago, Thorén, (1996, p. 389) argued that the clothing market is 'not adapted for people with unusual body dimensions and/or different kinds of functional impairments'. For instance, their figures do not fit into clothes made according to the sizing system used (this is a particular complaint by children with Down's Syndrome and their parents/carers). Further, body parts and joints may be affected by disability meaning certain clothing 'cuts' are not appropriate.

Fast-forward more than twenty years and mainstream clothing that is readily available on the high street is still not suitable for all types of bodies. This is despite a growing body of research that has explored the clothing needs and desires of disabled wearers (e.g. Chang et al., 2014; Stokes, 2010). Such research has found that different types of materials might restrict movement from devices such as crutches or braces (Reich and Otten, 1987). Further, individuals with sensory-related or skin conditions may have physical or behavioural reactions to different types of fabrics (Hilton et al., 2010), and conditions such as spina bifida and osteogenesis imperfecta can result in spinal curvature that affects how clothing fits (Kidd, 2006). There is an established body of literature citing a definite need for specially designed clothing to improve the disabled persons' comfort and confidence (e.g. Chang et al. 2009; Reich and Shannon, 1980; Stokes and Black, 2012). Attempts to improve clothing choices available for disabled wearers include adaptive apparel manufactured and marketed directly to disabled consumers as well as modification of mass-produced non-adaptive clothing (Carroll and Kincade, 2007). Questioning 'why is it difficult for disabled users to find suitable clothing?' Thorén (1996) discusses how, owing to the lack of suitable clothing available, sewing courses are given, especially for parents who want to learn how to make clothes for their disabled children. Further, clothing patterns for different kinds of impairment have been produced. For instance, patterns adapted to sitting in wheelchairs and unusually big or small bodies (Thorén, 1996).

Writing on creating special occasion garments for disabled young women, Kidd (2006, p. 161) argues that while young people 'loath to appear different', circumstances beyond their control, such as disability, may set them apart visually from their peers. Most ready-to-wear dresses have to be altered to accommodate the physical disability of young women. However, it is often difficult to adapt this clothing to create safe and functional garments, and as such the young people are often disappointed with the look and fit of the altered garment (Brandt, 1990). Kidd (2006) documents a creative design project involving the construction of custom-designed special occasion dresses for four women, aged 16–20, that have spina bifida or osteogenesis imperfecta. Common physical characteristics of these impairments include severe spinal curvature, resulting in extreme body asymmetry that requires the use of forearm crutches and waist-to-foot braces for upright mobility. Results of Kidd's (2006) research suggest that draping the muslin sample garments directly on the body is the most successful method of achieving good fit and creating the illusion of body symmetry and proportion. Some interesting design choices were made by the young people in the study, taking into consideration their disability. For instance, the participants chose long dresses to hide their leg braces. Further, sleeved styles were rejected because sleeves often got caught in their crutches as they walk.

Some disabled children and young people, to meet their physical needs, wear functional clothing, incorporating self-help and/or special-fit features. However, these garments and features are not unproblematic; clothing may not be desirable, due to distinct normative features. For instance, special features such as the use of Velcro, large zippers and other such features in unconventional places may be sources of stigma through differentiating the disabled from the non-disabled (Kaiser et al., 1985). Freeman et al., (1985) found that disabled people often feel stigmatised, even if special features on functional clothing are well hidden from others. The extent to which clothing is stigmatising depends on its difference from the norm.

Clothing and the management of identity

The manipulation of appearance symbols (such as clothing) affords opportunities for individuals to communicate visually to others that the disability is not the only aspect of the self (Kaiser et al., 1985). Due to its close (physically) relationship to the body, clothing can be a key area of manipulation by disabled people. Through the manipulation of clothing and accessories, the wearer may emphasise other aspects of self if concealment of the disability is not possible (Kaiser et al., 1985). Kaiser et al. (1985) found that strategies used by disabled young people include using clothes to conceal a disability, deflecting attention from a disability toward more normative but slightly discrediting attributes (e.g. dark glasses worn by those with a visual impairment), and compensation through fashionable dress. Other disabled young people in this study used dress to take advantage of their social uniqueness, through techniques such as wearing bright or prominent clothing.

Social competence and self-esteem for those with Down's Syndrome, with particular reference to clothes, is the focus of Rothschild's (1997) work. Children born with Down's Syndrome have numerous congenital defects. They do not attain normal height; their arms, legs, fingers and toes are short; their muscles lack tone or tension and the abdomen tends to be prominent. The face of the child is stigmatised by a large tongue, open mouth, saddle nose and epicanthal folds. Rothschild (1997) finds that clothing is an important cue in the formation of perceptions about the wearer. Further, clothing and the reaction elicited from others can be used to enhance self-esteem. Thus, identifying and understanding self-esteem builders such as clothing comfort may help the young people, their parents and educators find specific methods to inculcate positive behaviour.

Other research in this area has focussed on stigma and the lives of disabled children and their families (e.g. Green, 2003; Craig and Scambler, 2006; Werner and Shulman, 2015). The word stigma applies to any mark or attribute that sets some people apart from others and denotes a 'spoiled' social identity (Goffman, 1963). Goffman (1975) describes stigma as an attribute that casts deep discredit on the person who possesses it. Goffman (1975) presents three types of stigma: abominations of the body (various physical disfigurements), 'blemishes of individual character' (alcoholism, fascism etc.) and 'tribal' stigma (nationality, religion etc.) It is the first of these types of stigma that we are concerned with (see Pearson's discussion of street children, this volume, as an example of stigma related to 'blemishes of individual character'). Stigmatisation can occur at several levels, depending on the degree to which the body is blemished, and the character of the person discredited (Goffman, 1963). How clothing may be used to camouflage/hide or to detract attention from 'imperfections' is interesting. There has already been research revealing that camouflage can bring its own problems in relation to issues of identity (are people responding to the real me?), over-reliance on the camouflaged image in social interaction, and fears that the 'truth' will be discovered (see Coughlan and Clarke, 2002). Further, the extent to which clothing is stigmatising may depend on its difference from the norm, as in the case of a special zipper or other fasteners in unconventional places (Freeman et al., 1987). Other research has found that fashionable and attractive clothing is an effective means of appearance management by disabled children and young people, and is used to enhance social acceptance (Kaiser, 1997), and to foster feelings of belonging (Kidd, 2006).

In the ways discussed so far clothing can be considered as a way to disguise physical defects in which dressing becomes an 'act of deception' (Woodward, 2007, p. 125). When we dress we do so to make our bodies acceptable to a social situation (Entwistle, 2000). Clothes, then, are central to the performance, or 'curation', of our identities (Buse and Twigg, 2015, p. 1). Going further than this, Adam and Galinsky (2012, p. 919) in a discussion of 'enclothed cognition', posit that wearing clothes causes people to 'embody' the clothing and its symbolic meaning. Clothing has been identified as a sensual mediator between personal and social worlds, with the potential to provide insights into wearers' feelings, how they express identity, comprehend social mores, and prepare for social interaction (Candy and Goodacre, 2007). However, as Harvey (2007) reminds us, although clothes conceal, they may also emphasise what they conceal.

Disability, clothing and mobility

Whilst mobility is a clear focus of attention in disability studies and beyond, particularly in studies of wheelchair-bound children or those using walking aids (e.g. Noyes et al., 2017; Palisano et al., 2010), as well as visually impaired children (Worth, 2013), the relationship between clothing choices and individuals with mobility issues has not yet been adequately explored. Lamb (2001, p. 138) raised this argument, stating that most often the concern of disability scholars regarding mobility is with the built environment, as opposed to the 'near, portable environment of dress'. There are some exceptions, however. Kratz et al. (1997) discuss wheelchair users' experiences of non-adapted and adapted clothing during sailing, quad rugby or wheel-walking. Kratz et al. (1997) found that wheelchair users associated significantly greater comfort with the use of adapted clothes. Further, the wheelchair users set a higher priority upon independence in work or leisure activities than in activities of daily living. The results of the study confirm the value of adapting sportswear for disabled wearers. Further, in Kidd's (2006) previously mentioned study into the construction of custom-designed special occasion dresses for young women with spina bifida or osteogenesis imperfecta, the author found that sleeved styles of dresses were rejected because sleeves often got caught in their crutches as they walk. This demonstrates the importance of considering mobility and movement in the design of clothing with disabled children.

Nicholson et al. (2001) assess the upper-limb function and movement in children with cerebral palsy wearing Lycra garments. The authors found that Lycra garments are helpful for some children with cerebral palsy, for instance in improving stability in sitting and in smoothness of arm movements. However, children in the study had some difficulties when wearing the garments, including problems with toileting. Results from Nicholson et al.'s (2001) research suggest that the functional benefit of Lycra garments for children with cerebral palsy is mainly due to improvements in proximal stability, but this should be weighed against the inconvenience and loss of independence. With attention to ability, adaptation and engagement, Kabel et al. (2016) consider clothing-related participation barriers. The authors find that lack of adaptive or appropriate clothing for disabled people can become a barrier, preventing engagement in meaningful activities. Ultimately, this has negative implications for rehabilitation. While there has been an emphasis on the wearing of clothes in social situations in a more static (stationary) sense, there is more work to be done on clothing and im/mobility.

Dress and undressing

Independent dressing skills by disabled children and young people has also been the focus of some research (Hughes et al. 1993; Young et al. 1986), as has the putting on and removing of clothes, particularly coats and jackets (e.g. Reese and Snell, 1991). This area of scholarship is important when considering that dressing difficulties of disabled persons has emerged as a concern for rehabilitation (Lamb, 2001), whilst learning to dress has long been considered a fundamental skill toward independence for disabled people (Levitan-Rheingold, 1980). In recognition of this, a learn-to-dress storybook has been developed in conjunction with a practical and functional children's clothing range, to aid children with autism (see Moosa, 2010). This is an excellent example of impactful research and of the importance of co-production.

For some disabled children, such as those with visual impairments, dressing is a difficult and sometimes impossible task (Sudha and Bhawana, 2011). Chitora (2011) finds that visually impaired children face problems such as zipping up a coat, tying their shoe laces or buttoning their shirts, as well as identifying the fabric and colour of clothing. Front opening garments were preferred by participants in Chitora's (2011) study, due to ease in fastening. This also helped respondents to distinguish the front from the back of the garment. Whilst attributes such as comfort, durability and style affected the buying of clothing, most decisions about clothing purchase were made by respondent's family members.

A related yet distinct concern is decisions around *un*dressing or 'disrobing' as it is more frequently called in the field of disability studies. Stokes and Black (2012) assess the clothing needs of young disabled girls. In doing so, they pay attention to the functional, expressive and aesthetic consumer needs. While functional considerations were most often reported, adolescents also indicated a number of expressive and aesthetic considerations. Regardless of their level of clothing interest, the common functional considerations identified were issues with fit and difficulty 'donning and doffing' (putting on and taking off clothing), including trouble with garment fasteners (Stokes and Black, 2012, p. 179). Carlson et al. (2008) discuss public disrobing in two disabled children who disrobed in school settings. The children also urinated in their clothing in order to gain access to new and more preferred clothing. An intervention gave the children a choice to change into their preferred clothes at scheduled opportunities during the day. This research found that, by scheduling opportunities to change clothes, the intervention lessened each child's motivation to disrobe and decreased incidents of urinary incontinence with both children. Garments have also been designed which restrict unassisted disrobing (e.g. see Royal, 1993). Research into the disrobing practices of disabled children and young people is scant and is an area worthy of further academic attention.

Conclusion

This chapter is concerned with disabled children and young people's experience of choosing and wearing clothing, and how they use clothing to express identity and prepare for social interaction. We highlighted four key thematic areas in the literature: finding appropriate clothing; clothing and the management of identity; disability, clothing and mobility; and dress and undressing. This chapter has highlighted that, for some disabled children and young people, mainstream clothing is not appropriate for their needs. As such, they turn to functional clothing with self-help and special-fit features. However, some special features on functional clothing can lead to feelings of stigmatisation if they are not well hidden from others. Other research reviewed found that fashionable and attractive clothing is an effective means of appearance management by disabled children and young people, and is used to enhance social acceptance (Kaiser, 1997) and to foster feelings of belonging (Kidd, 2006). Importantly, however, the aesthetics of clothing need to be weighed up against their functional aspects, as research reviewed in this chapter found that lack of adaptive or appropriate clothing for disabled children and young people can prevent engagement in meaningful activities (Kabel et al. 2016). Ultimately, this has negative implications for rehabilitation. It is clear that rehabilitation professionals need to take clothing and apparel-related issues faced by disabled children seriously, so that opportunities for social participation can be maximised.

The preoccupation in the geography literature has been the multiple physical barriers in the built environment that impede children and young people's participation in different spaces and places. Researchers in other disciplines have begun to consider the extent to which items of clothing may contribute to physical and social exclusion for disabled children (for instance through considerations of normalising techniques used by disabled children and young people, as discussed throughout this chapter, such as: 'making do' with ready-to-wear clothing through resourceful adaptations; use of clothing to conceal a disability; and deflection of attention towards other aspects of the self). However, whilst there has been an emphasis on the wearing of clothes in social situations in a more static (stationary) sense, there is a relative deficit of research on dressing and undressing and clothing and im/mobility. This is an important area for future research – after all, the more we understand about how children are permitted, or restricted, to move in, through, and beyond spaces and places, by their clothing, the more we can work towards assisting their inclusion in these spaces.

Notes

1 Spina bifida is a birth defect that occurs when the spinal cord, brain, or protective coverings for the spinal cord or brain do not develop completely.
2 Osteogenesis imperfecta, often referred to as 'brittle bone disease', is a genetic bone disorder.
3 An idea of the self, constructed from the beliefs one holds about oneself and the responses of others.

References

Adam, H. and Galinsky, A. D. (2012). Enclothed cognition. *Journal of Experimental Social Psychology*. 48. (4). pp. 918–925.

Boden, S. (2006). Dedicated followers of fashion? The influence of popular culture on children's social identities. *Media, Culture & Society*. 28. (2). pp. 289–298

Brandt, L. (1990). Designing for the disabled. *Family Circle*. 15. pp. 18–19.

Buse, C.E. and Twigg, J. (2015). Clothing, embodied identity and dementia: Maintaining the self through dress. *Age, Culture, Humanities*. 2. http://agecultcurehuma nities.org/WP/clothing-embodied-identity-and-dementia-maintaining-th e-self-through-dress/ (last accessed 20 Jan 2019).

Candy, F.J. and Goodacre, L. (2007). The wardrobe and well being: Exploring relationships between women living with rheumatoid arthritis and their clothing. Paper presented at Helen Hamlyn Centre, London, RCA Include Conference, 2–4 April. www.a-brand.co.uk/pdfs/ab_paperforinclude.pdf (last accessed 20 Jan 2019).

Carlson, J. I., Luiselli, J. K., Slyman, A., & Markowski, A. (2008). Choice-making as intervention for public disrobing in children with developmental disabilities. *Journal of Positive Behavior Interventions*. 10. (2). pp. 86–90.

Carroll, K. E. and Kincade, D. H. (2007). Inclusive design in apparel product development for working women with physical disabilities. *Family & Consumer Sciences Research Journal*. 35. (4). pp. 289–315.

Chang, H. J., Hodges, N., & Yurchisin, J. (2014). Consumers with disabilities: A qualitative exploration of clothing selection and use among female college students. *Clothing and Textiles Research Journal*. 32. (1). pp. 34–48.

Chang, W. M., Zhao, Y. X., Guo, R. P., Wang, Q. and Gu, X. D. (2009). Design and study of clothing structure for people with limb disabilities. *Journal of Fiber Bioengineering and Informatics*. 2. (1). pp. 62–67.

Chitora, B. (2011). *A Study on Clothing Practices of Visually Impaired Children*. Doctoral dissertation, Udaipur, MPUAT.

Connors, C., and Stalker, K. (2007). Children's experiences of disability: Pointers to a social model of childhood disability. *Disability & Society*. 22. (1). pp. 19–33.

Cook, D. (2004). *The Commodification of Childhood*. Durham, NC: Duke University Press.

Corker, M. and Davis, J. M. (2000). Disabled children: Invisible under the law. In J. Cooper and S. Vernon (ed.). *Disability and the Law*. London: Jessica Kingsley.

Coughlan, G. and Clarke, A. (2002). Shame & Burns. In P. Gilbert and J. Miles (eds.) *Body Shame*. Hove: Brunner-Routledge. pp. 155–170.

Craig, G. M. and Scambler, G. (2006). Negotiating mothering against the odds: Gastrostomy tube feeding, stigma, governmentality and disabled children. *Social Science & Medicine*. 62. (5). pp. 1115–1125.

Crewe, L. and Collins, P. (2006). Commodifying children:Fashion, space, and the production of the profitable child. *Environment and Planning A*. 38. (1). pp. 7–24.

Davis, J., Watson, N. and Cunningham-Burley, S. (2008). Disabled children, ethnography and unspoken understandings: The collaborative construction of diverse identities. In P.M. Christensen and A. James (eds.) *Research with Children: Perspectives and Practices*. New York: Routledge. pp. 220–238.

Entwistle, J. (2000). Fashion and the fleshy body: Dress as embodied practice. *Fashion Theory*. 4. (3). pp. 323–347.

Freeman, C.M., Kaiser, S.B. and Wingate, S.B. (1985). Perceptions of functional clothing by persons with physical disabilities: A social-cognitive framework. *Clothing and Textiles Research Journal*. 4. (1). pp. 46–52.

Freeman, C. M., Kaiser, S. B., & Chandler, J. L. (1987). Perceptions of functional clothing by able-bodied people: The other side. *Journal of Consumer Studies & Home Economics*. 11. (4). pp. 345–358.

Goffman, E. (1963). *Stigma: Notes on the Management of Spoiled Identity*. New York: Simon and Schuster.

Goffman, E. (1975). *Stigma: The Social Functions of Disabilities*. Paris: Minuit.

Goodley, D., and Runswick-Cole, K. (2010). Len Barton, inclusion and critical disability studies: Theorising disabled childhoods. *International Studies in Sociology of Education*. 20. (4). pp. 273–290.

Goodley, D., Runswick-Cole, K., and Liddiard, K. (2016). The dishuman child. *Discourse: Studies in the Cultural Politics of Education*. 37. (5). pp. 770–784.

Green, S. E. (2003). 'What do you mean "what's wrong with her?"': stigma and the lives of families of children with disabilities. *Social Science & Medicine*. 57. (8). pp. 1361–1374.

Harvey, J. (2007). Showing and hiding: Equivocation in the relations of body and dress. *Fashion Theory*. 11. (1). pp. 265–294.

Hilton, C. L., Harper, R. H., Kueker, A. R. et al. (2010). Sensory responsiveness as a predictor of social severity in children with high functioning autism spectrum disorders. *Journal of Autism and Developmental Disorders*. 40. (8). pp. 937–945.

Hodge, N. and Runswick-Cole, K. (2013). 'They never pass me the ball': Exposing ableism through the leisure experiences of disabled children, young people and their families. *Children's Geographies*. 11. (3). pp. 311–325.

Holt, L. (2004). Children with mind-body differences: Performing disability in primary school classrooms. *Children's Geographies*. 2. (2). pp. 219–236.

Holt, L. (2006). Exploring 'other' childhoods through quantitative secondary analyses of large scales surveys: Opportunities and challenges for children's geographers. *Children's Geographies*. 4. (2). pp. 143–155.

Holt, L. (2010). Young people's embodied social capital and performing disability. *Children's Geographies*. 8. (1). pp. 25–37.

Hughes, M. W., Schuster, J. W. and Nelson, C. M. (1993). The acquisition of independent dressing skills by students with multiple disabilities. *Journal of Developmental and Physical Disabilities*. 5. (3). pp. 233–252.

Imrie, R. and Edwards, C. (2007). The geographies of disability: Reflections on the development of a sub-discipline. *Geography Compass*. 1. (3). pp. 623–640.

Kabel, A., McBee-Black, K. and Dimka, J. (2016). Apparel-related participation barriers: Ability, adaptation and engagement. *Disability and Rehabilitation*. 38. (22). pp. 2184–2192.

Kaiser, S.B. (1997). *The Social Psychology of Clothing*. New York: Fairchild.

Kaiser, S. B., Freeman, C. M., and Wingate, S. B. (1985). Stigmata and negotiated outcomes: Management of appearance by persons with physical disabilities. *Deviant Behavior*. 6. (2). pp. 205–224.

Kellegrew, D. H. (2000). Constructing daily routines: A qualitative examination of mothers with young children with disabilities. *American Journal of Occupational Therapy*. 54. (3). pp. 252–259.

Kelly, B., and Byrne, B. (2015). Valuing disabled children and young people. *Child Care in Practice*. 21. (1). pp. 1–5.

Kidd, L. K. (2006). A case study: Creating special occasion garments for young women with special needs. *Clothing and Textiles Research Journal*. 24. (2). pp. 161–172.

Kratz, G., Söderback, I., Guidetti, S., Hultling, C., Rykatkin, T. and Söderström, M. (1997). Wheelchair users' experience of non-adapted and adapted clothes during sailing, quad rugby or wheel–walking. *Disability and Rehabilitation*. 19. (1). pp. 26–34.

Lamb, J. M. (2001). Disability and the social importance of appearance. *Clothing and Textiles Research Journal*. 19. (3). pp. 134–143.

Levitan-Rheingold, N. (1980). Learning to dress: A fundamental skill toward independence for the disabled. *Rehabilitation Literature*. 41. (3–4). pp. 72–75.

Lindsay, S., and McPherson, A. C. (2012). Experiences of social exclusion and bullying at school among children and youth with cerebral palsy. *Disability and Rehabilitation*. 34. (2). pp. 101–109.

Moosa, N. (2010). *A Learn-to-Dress Storybook in Conjunction with a Practical and Functional Children's Wear Range: To aid children with autism*. Doctoral dissertation, Cape Town, Cape Peninsula University of Technology.

Nicholson, J. H., Morton, R. E., Attfield, S. and Rennie, D. (2001). Assessment of upper-limb function and movement in children with cerebral palsy wearing lycra garments. *Developmental Medicine and Child Neurology*. 43. (6). pp. 384–391.

Noyes, J., Spencer, L. H., Bray, N., Kubis, H. P., Hastings, R. P., Jackson, M., and O'Brien, T. D. (2017). Conceptualization of physical exercise and keeping fit by child wheelchair users and their parents. *Journal of Advanced Nursing*. 73. (5). pp. 1111–1123.

Oliver, M. (1996). *Understanding Disability: From theory to practice*. London: Macmillan.

Palisano, R. J., Hanna, S. E., Rosenbaum, P. L., and Tieman, B. (2010). Probability of walking, wheeled mobility, and assisted mobility in children and adolescents with cerebral palsy. *Developmental Medicine & Child Neurology*. 52. (1). pp. 66–71.

Parr, H. and Butler, R. (1999). New geographies of illness, impairment and disability. In R. Butler and H. Parr (eds.). *Mind and Body Spaces: Geographies of illness, impairment and disability*. London/New York: Routledge.

Piacentini, M., and Mailer, G. (2004). Symbolic consumption in teenagers' clothing choices. *Journal of Consumer Behaviour: An International Research Review*. 3. (3). pp. 251–262.

Pickering, D. (2018). The embodiment of disabled children and young people's voices about participating in recreational activities: Shared perspectives. In K. Runswick-Cole, T. Curran and K. Liddiard (eds.). *The Palgrave Handbook of Disabled Children's Childhood Studies* (pp. 101–123). London: Palgrave Macmillan.

Pilcher, J. (2010). What not to wear? Girls, clothing and 'showing' the body. *Children & Society*. 24. (6). pp. 461–470.

Pyer, M., Horton, J., Tucker, F., Ryan, S. and Kraftl, P. (2010). Children, young people and 'disability': Challenging children's geographies? *Children's Geographies*. 8. (1). pp. 1–8.

Ratna Jyothi, D. (1988). *Dress Designing for Disabled*. Doctoral dissertation, Rajendranagar, Hyderabad: Acharya Charya NG Ranga Agricultural University.

Reese, G. M., and Snell, M. E. (1991). Putting on and removing coats and jackets: The acquisition and maintenance of skills by children with severe multiple disabilities. *Education and Training in Mental Retardation*. 26. (4). pp. 398–410.

Reich, N., and Shannon, E. (1980). Handicap: Common physical limitations and clothing-related needs. *Home Economics Research Journal*. 8. (6). pp. 437–444.

Reich, N. and Otten, P. (1987). What to wear: A challenge for disabled elders. *The American Journal of Nursing*. 87. (2). pp. 207–210.

Rosenbaum, P. and Gorter, J. W. (2011). The 'F-words' in childhood disability: I swear this is how we should think! *Child: Care, Health and Development*. 38. (4). pp. 457–463.

Ross, J. and Harradine, R. (2004). I'm not wearing that! Branding and young children. *Journal of Fashion Marketing and Management: An International Journal*. 8. (1). pp. 11–26.

Rothschild, M. (1997). *Living in the Community (Down's Syndrome): Social competence and social self-esteem with particular reference to clothes*. Unpublished thesis, Loughborough: Loughborough University.

Royal, P. (1993). Garment, particularly for restricting wearer's undressing. U.S. Patent 5,208,918.

Runswick-Cole, K., Goodley, D., and Lawthom, R. (2018). Resilience in the lives of disabled children: A many splendoured thing. In K. Runswick-Cole, T. Curran and K. Liddiard (eds.). *The Palgrave Handbook of Disabled Children's Childhood Studies* (pp. 425–442). London: Palgrave Macmillan.

Rysst, M. (2010). 'I am only ten years old': Femininities, clothing-fashion codes and the intergenerational gap of interpretation of young girls' clothes. *Childhood*. 17. (1). pp. 76–93.

Saunders, K. (2000). *Happy Ever Afters: A storybook guide to teaching children about disability*. London: Trenton Books.

Shakespeare, T. and Watson, N. (2001). The social model of disability: An outdated ideology? In S. Barnatt and B. Altman (eds.) *Exploring Theories and Expanding Methodologies: Where we are and where we need to go*. Bingley: Emerald Group Publishing. pp. 9–28.

Singh, V. and Ghai, A. (2009). Notions of self: Lived realities of children with disabilities. *Disability & Society*. 24. (2). pp. 129–145.

Stokes, B. M. (2010). *Clothing Needs of Teen Girls with Disabilities*. Doctoral dissertation, Pullman, WA: Washington State University.

Stokes, B., and Black, C. (2012). Application of the functional, expressive and aesthetic consumer needs model: Assessing the clothing needs of adolescent girls with disabilities. *International Journal of Fashion Design, Technology and Education*. 5. (3). pp. 179–186.

Sudha, B., and Bhawana, C. (2011). Clothing practices of visually impaired children. *Asian Journal of Home Science*. 6. (1). pp. 39–42.

Thorén, M. (1996). Systems approach to clothing for disabled users. Why is it difficult for disabled users to find suitable clothing. *Applied Ergonomics*. 27. (6). pp. 389–396.

Torrell, V. B. (2004). Adults and children debating sexy girls' clothes. In H. Brembeck, B. Johansson, and J. Kampmann (eds). *Beyond the Competent Child*. Fredriksberg: Roskilde University Press. pp. 251–274.

Turnbull, A. P., Pereira, L., and Blue-Banning, M. J. (1999). Parents' facilitation of friendships between their children with a disability and friends without a disability. *Journal of the Association for Persons with Severe Handicaps*. 24. (2). pp. 85–99.

Valentine, G. (2000). Exploring children and young people's narratives of identity. *Geoforum*. 31. (2). pp. 257–267.

Valentine, G. and Skelton, T. (2003). Living on the edge: The marginalisation and resistance of D/deaf youth. *Environment and Planning A*. 35. (2). pp. 301–321.

Valentine, G. and Skelton, T. (2007). Re-defining norms: D/deaf young people's transitions to independence. *Sociological Review*. 55. (1). pp. 104–123.

von Benzon, N. (2010). Moving on from ramps? The utility of the social model of disability for facilitating experiences of nature for disabled children. *Disability & Society*. 25. (5). pp. 617–626.

von Benzon, N. (2017a). Confessions of an inadequate researcher: Space and supervision in research with learning disabled children. *Social and Cultural Geography*. 18. (7). pp. 1039–1058.

von Benzon, N. (2017b). Unruly children in unbounded spaces: School-based nature experiences for urban learning disabled young people in Greater Manchester, UK. *Journal of Rural Studies*. 51. pp. 240–250.

Watson, A., Blanco, J., Hunt-Hurst, P. and Medvedev, K. (2010). Caregivers' perceptions of clothing for people with severe and profound intellectual disabilities. *Perceptual and Motor Skills*. 110. (3). pp. 961–964.

Werner, S. and Shulman, C. (2015). Does type of disability make a difference in affiliate stigma among family caregivers of individuals with autism, intellectual disability or physical disability? *Journal of Intellectual Disability Research*. 59. (3). pp. 272–283.

Wickenden, M. and Kembhavi-Tam, G. (2014). 'Ask us too!' Doing participatory research with disabled children in the global south. *Childhood*. 21. (3). pp. 400–417.

Woodward, S. (2007). *Why Women Wear What They Wear*. New York: Berg Publishers.

Worth, N. (2008). The significance of the personal within disability geography. *Area*. 40. (3). pp. 306–314.

Worth, N. (2013). Visual impairments in the city: Young people's social strategies for independent mobility. *Urban Studies*. 50. (3). pp. 574–586.

Young, K. R., West, R. P., Howard, V. F. and Whitney, R. (1986). Acquisition, fluency training, generalization, and maintenance of dressing skills of two developmentally disabled children. *Education and Treatment of Children*. 9. pp. 16–29.

3 'They should have stayed'

Blaming street children and disruption of the intergenerational contract

Gemma Pearson

Chapter summary

Street-living children in Tanzania find themselves excluded from mainstream society due to the stigmatisation they receive for their social transgressions. Belcher and DeForge (2012) explain that the stigmatising and othering of homeless populations is, in part, a means of placing the blame for homelessness on the individual, absolving broader society of responsibility for the structural factors that cause inequality. Street-living children in Tanzanian society transgress social norms by appearing to live independently of their parents and wider family in a context where intergenerational reciprocity is paramount. Former research on street-living children has highlighted the importance of children and young people's relationships with peers as sources of solace, cooperation and home-making on the street (Beazley, 2003; van Blerk, 2012). Although important, I argue that these street-based relationships are systemically different from their former relationships with their families and home communities, which are formed around informal intergenerational contracts, and cannot substitute the institutional stability offered by familial ties and association. Within the context of family breakdown, this chapter will draw on the concept of intergenerational reciprocity, street child literature and primary research to explore who is responsible for street-living children, how they are responsible and to what extent.

Introduction

Street-living children[1] in Tanzania find themselves excluded from mainstream society due to the stigmatisation they receive for their dual transgressions: being visible on the street, and therefore not with their families (Davies, 2008), and conducting private activities in public space, such as eating and sleeping (Wells, 2015). Belcher and DeForge (2012) explain that the stigmatising and othering of homeless populations is, in part, a means of placing the blame for homelessness on the individual, absolving broader society of responsibility for the structural factors that cause inequality. This blame is justified by dehumanising those who are homeless; labelling them to separate them from broader society and depicting them as 'threatening (dangerous), nonproductive, and personally culpable' (Belcher and DeForge, 2012: 931). The anomaly of street-living children, due to their social transgressions, can elicit mistreatment and a lack of respect for their rights. Street-living children in Tanzanian society also transgress social norms by appearing to live independently of their parents and wider family in a context where intergenerational reciprocity is paramount (Evans and Becker, 2009; Twum-Danso, 2009 in Ghana; Day and Evans, 2015 in southern Africa). The intergenerational contract is an unwritten agreement which obligates children to perform tasks asked of them by their parents in return for care and protection (van Blerk, 2012). This arrangement can breakdown when either party believes the other to not be upholding the implicit terms of the agreement; as in cases of abuse or neglect on the side of the parent, or disobedience from the child. In many cases, street-living children who have found homelife unbearable step towards living individualistically as a survival mechanism for escaping dangerous situations (at home and/or on the street) and providing for their own needs. In a collectivist society, however, individualism is a cause for suspicion.

Former research on street-living children has highlighted the importance of children and young people's relationships with peers as sources of solace, cooperation and home-making on the street (Beazley, 2003; van Blerk, 2012). Although important, I argue that these street-based relationships are systemically different from their former relationships with their families and home communities, which are formed around informal intergenerational contracts, and cannot substitute the institutional stability offered by familial ties and association. It is necessary to note that populations of street-living children differ across locations with regard to their connections to their families and communities of origin. Factors that may influence these differences include distance and proximity to their families (van Blerk, 2012) and children's motives for moving to the street; if moving for economic reasons, children may maintain closer ties to their families through remittances (Hecht, 1998). This chapter relates to research with street-living children and young people in northern Tanzania[2] where, for the majority, children reported to have left home due to circumstances of abuse, neglect or a breakdown of relationships among the family. Children and young people involved in this study rarely reported returning home to visit their families. They more often discussed running away from family members and engaging in movement as a mechanism to avoid being returned to their communities of origin. This chapter does not speak to all populations of street-living children, but specifically to those where children have become, or are in the process of becoming, estranged from their families in a context where concepts of intergenerational contracts provide the predominant framework of social support.

Within this context of family breakdown, this chapter will draw on the concept of intergenerational reciprocity, street child literature and primary research to explore who is responsible for street-living children, how they are responsible and to what extent. Services that effectively emulate the responsibility that families have for investing in children's futures, as well as understanding children's wishes and desires, are important for buffering the hardships street-living children and young people face on the street so that children can invest in activities beyond daily survival. However, this chapter will argue that family represents a crucial support network for children and young people in Tanzania which is not easily replicated by street-based relationships and networks.

Research context and methodology

This chapter draws on nine months of fieldwork in northern Tanzania between 2014 and 2016. Using grounded theory methodology (GTM), semi-structured interviews were carried out with 55 street-living children and young people, former street-living children, social workers and community members, as well as participant observation of children on the street in two field sites in northern Tanzania. Working with local researchers, and former social workers as gate keepers, group interviews were conducted with children and young people in various street locations across the field sites. Consent was obtained verbally and interviews were recorded with participants' permission. The initial purpose of the research was to understand the role that relationships and social networks play in children's wellbeing. Using GTM (Glaser and Strauss, 1967), interviews were conducted and analysed concurrently to identify new lines of enquiry and clarify emerging findings, and participants, street-living populations and non-street-living populations were sampled theoretically as data categories developed. The data indicated that street-living children and young people are taking responsibility for their own needs. This core concern permeated the other data categories and it became apparent that many of the children and young people in this study desired a responsible adult to take care of their needs and advocate on their behalf; this was typically framed with a view to their long-term prospects (see Boampong, this volume, for examples of this in a Ghanaian context).

Street-living children and intergenerational contracts of responsibility

Street-living children and young people are far from isolated and, in fact, build intricate and varied social networks on the street. However, street relationships differ from children and young people's relationships with their families due to expectations of responsibility inherent in the familial unit (see Willman, this volume, for a discussion of expectations of responsibility within a family context). Therefore, this chapter argues that responsibility is a powerful lens through which to analyse street living children's relationships. When children leave home to live on the street they are stepping away from a family and community who are, at least nominally, responsible for their care. Children and young people can build new relationships on the street that provide company, assistance and cooperation. It is rare, however, for someone who a child recently met on the street to take responsibility for the child's welfare to the extent that is expected from the family. In a country with under-resourced social service provision, such as Tanzania, when children leave home they find themselves entirely responsible for meeting their own needs and shoulder the burden of stress and anxiety that accompanies this responsibility. Children in Tanzania leave home for a variety of reasons, including: parental separation and subsequent disagreements over their custody, verbal or physical abuse from a family member, step-parent or teacher, death of a parent and disagreements over inheritance, and a real, or perceived, lack of income opportunities at home.

In Tanzania, and many parts of sub-Saharan Africa, society expects that children are responsible to and for their families through notions of inter-generational contracts (van Blerk and Ansell, 2007) and reciprocity (Twum-Danso, 2009). van Blerk and Ansell explain, in the context of southern Africa:

> Inter-generational contracts revolve around two different kinds of dependency separated in time: the dependency associated with childhood and the dependency associated with old age [...] Within families, parents carry out their obligations in anticipation of future returns, children in recognition of past benefits. (2007, p. 870)

Children's responsibilities to their families are actualised through their work around the home (Libório and Ungar, 2010), obedience (Hollos, 2002), remittances (Hecht, 1998 and his concept of 'nurturing children' vis a vis 'nurtured children') and caring for elderly family members in later life. For this reason, children who leave home are not only taking responsibility for themselves, but they are also communicating to others that they may have abandoned some or all of their responsibilities to their families, intentional or otherwise. This apparent rejection of their responsibilities to their family and community is one reason why children and young people living on the street face stigmatisation. As Awedoba argues, relationships of reciprocity between African parents and their children are ingrained in concepts of moral obligation:

> Society does not spare those African parents and children who fail in their reciprocal obligations. The recalcitrant child or parent may be ridiculed or gossiped about by concerned others. The aged parent may curse the negligent child who neglects that aged parent. (2002, p. 90)

A fundamental misunderstanding among broader society of the reasons why children have left home imbues children and young people with blame, for abandoning their responsibilities, and casts doubt on their character. As one social worker explains:

> And this assumption [is] always, like, a child who runs away from home is a naughty child, doesn't listen to their parents, is sort [of] disobedient. (Fred, social worker)

Similarly, the assumption that street-living children are disobedient and stubborn is reiterated by others in the community:

> Tanzanians think the street children are the stubborn children from the home, they are people who don't want to hear anything from their parents, they don't want to listen [to] anything [their parents say], they are rude, they are those people who don't want to listen to the parents, listen to the community, they don't want to be punished. Stubborn ones. (Irene[3], 26, female teacher)

Children who live on the street are imbued with certain character traits: rude, stubborn and disobedient. Such children may be considered as acting within their own interest without due consideration of the best interest of their parents, families or communities of origin. Acting in one's own interest is something which, arguably, goes against core African values, as the African Charter on Human and People's Rights, Article 29, stipulates:

> The individual shall also have the duty: 1. To preserve the harmonious development of the family and to work for the cohesion and respect of the family; to respect his parents at all times, to maintain them in case of need (African Commission on Human and Peoples' Rights, 1981)

Similar sentiments are also extended in the African Charter on the Rights and Welfare of the Child (ACRWC), Article 31:

> Children have responsibilities towards their families and societies, to respect their parents, superiors and elders, to preserve and strengthen African cultural values in their relation with other members of their communities. (Organisation of African Unity, 1990)

To act in an individualist manner, i.e. leaving home and living independently on the street, attracts suspicion and scorn from others. This suspicion arises, in part, from the imagined or assumed negative character traits of a child that would precede family estrangement, as one young person explains:

> We were brought to prison, I was brought with my friend [...]. His mother hesitated to pay my bail saying that 'I have met his mother and told her that he is at the police [station] and [his] mother has not come. I can't pay his bail, you never [know] what he has already done to wrong his mother.' That means that I am not only hated by [my] parents but also with the society surrounding me. (Pauli, age 22, male).

In this example, the breakdown of relationship between Pauli and his mother means that his friend's mother does not trust him. Pauli extends this example to the wider community, explaining that the rejection from this family leads others in society to reject him due to a suspicion that he has wronged his parents and, therefore, has a questionable character. This example also illustrates the difficulties street-living children and young people face building relationships of trust with other adults when they are knowingly, or assumed to be, estranged from their family. Particularly, the reason the woman gave for not paying Pauli's bail was due to the poor relationship he had with his mother, rather than concerns over the transgressions that led him to be taken into police custody.

Within familial relationships in Tanzania there are consistent expectations of responsibility in the form of intergenerational contracts, expressed as 'duty' in the African Commission on Human and People's Rights (1981), and family members suffer 'social sanctions' when these responsibilities are left unfulfilled (Lassen and Lilleør, 2008). Therefore, I argue that street-living children and young people in Tanzania suffer stigmatisation not only due to their association with the street but also because of their perceived abnegation of the intergenerational contract.

Dirty bad-boys: why street children struggle for employment

Street-living children in Tanzania are heavily stigmatised for their assumed lack of morals associated with their unsightly appearance and unseemly behaviour, as expressed by participants in my study.

> Jamil[4]: You may find yourself looking kind of dirty, your clothes are undefined, you don't get along with people you live alongside [wider community] in terms of the language you are using: such behaviours makes you be recognised as a street child.
> RA[5]: Mmm. For example, which behaviours?
> Kelvin[6]: Gambling [...], sitting at the railway, smoking marijuana, cigarettes and so forth.

Living on the street and involving oneself in smoking, drug use, alcohol consumption and theft fosters mistrust between street-living children and the wider community. Due to a scarcity of income-generating opportunities, many resort to theft or other illegal activities to earn enough money to meet their needs (Shand et al., 2017). Such livelihood strategies exacerbate the stigma children and young people receive since it confirms communities' stereotypes of street-living children as individuals with questionable moral standing. Due to their challenging experiences, many street-living children find it difficult to trust adults (Conticini, 2008) and find it incredibly difficult to live up to adults' behavioural expectations. Indeed, some defy adults' expectations by rejecting societal norms in reaction to the exclusion they themselves may have experienced (Beazley, 2003).

The stigmatisation that street-living children face in Tanzania is multidimensional. One research participant, a British charity worker who had worked in Tanzania for over 30 years, explained that street-living youth are perceived as 'dirty, bad-boys' and that for a child to end up on the street was a parent's worst fear. A social worker expressed that community members often do not understand how a child could leave home to live on the street, instead taking the attitude that 'they [the children] should have stayed [at home]'. He explained that all children are beaten at home and that this is a crucial part of discipline, so there was little empathy for those who leave because of 'abuse': he suggested that because corporal punishment was so common in Tanzania, at home and in school (Hecker et al., 2014), many find it difficult to appreciate the extreme levels of physical abuse that some children experience. Furthermore, Frankenberg et al. (2010) investigated discipline strategies in Tanzania and identified that, for their research participants, to not beat one's child was tantamount to not caring about their development and risked them becoming irresponsible and disrespectful. Community members' suggestions that street-living children are those who do not want to listen to their parents or be disciplined (see previous section) implies that leaving home was due to the child's character flaw, rather than a fault of the parents. Young people themselves conceded that they viewed their beatings at home differently with hindsight:

> Kasim:[7] A lot of people before growing up, when you are still at home and given a chore to perform [...]. maybe you have done a small mistake and you don't see the mistake and then you are told that you have done that mistake and it reaches a stage father or mother decides to punish you, you feel like you are not treated right. But once you have grown up and reflect back you see that a certain day I made a mistake but back then you felt that you were treated unfairly, but it is not that way: parents love us, they cannot beat us if you have not done wrong, they cannot give you chores just to spite you.

Kasim looks back on his parents' beatings as an expression of their love, orientating his younger self as ignorant to their intentions by his sense of injustice at the time. By doing this Kasim is accepting that he was in the wrong to resist his parents' physical punishment and avoid being taught that he had made a mistake, thereby appropriating the blame that others may have placed on him for leaving home.

In Tanzania, the family provide a frame of reference for a child, advocate for them and can vouch for their trustworthiness: without this frame of reference community members may find it difficult to trust street-living children and young people. Some of the young people in my study went so far as to suggest that it was impossible to gain meaningful employment – the sort of job that might provide reliable income and opportunities – without the support of a referee who could vouch for their reliability and mediate their relationship with a new employer; something a parent would ordinarily do. Unfortunately, as one young person explains, it is difficult for street-living children and young people to find an adult who will trust them enough to play the role of referee:

Pauli[8]: When you go to ask for a job somewhere, if you are asked to bring someone who can stand before [be a referee], you can bring your parent, [but] where will I go to start? [Go to] someone I don't speak to [and say] 'Mother[9] lets go, and you pay a bill [another term for 'referee' or 'providing a reference'] for me somewhere'? And someone else, even if he loves you, he cannot stand before you because he does not know when or at what time you might betray [him].

In the excerpt above, Pauli is suggesting that even if he finds an adult who cares about him, they may not be willing to take the reputational risk of recommending him to an employer in case Pauli betrays their trust by acting inappropriately and negatively impacting the relationship between the referee and the employer. A study into street children's transitions into adulthood in northern Tanzania similarly found that in some cases NGOs faced difficulties matching young people with apprenticeships due to the stigma that street-involved populations face, and concerns, sometimes validated, from business owners that apprentices would steal from them (Blackford et al., 2008). Blackford et al. add that, in instances where young people do violate the terms of their apprenticeship, they adversely impact opportunities for other youth 'by putting into question the integrity and credibility of a center that vouched for them' (2008: 25). Years of meeting their needs via theft make it extremely difficult for street-living young people to access the necessary opportunities to become a self-sufficient adult. This example signifies that, although older street-living children and young people are able to meet their needs for daily survival, they may require a responsible adult who can support them in providing for their futures.

Relationships of reciprocity on the street

Relationships of reciprocity are transposed onto, and generated among, children's encounters on the street. Among my research participants, and converse to Pauli's experiences above, those who formed close relationships with adults in the market-place were offered more opportunities for work and protection. These relationships were often mutually beneficial as children would perform chores for vendors, motorbike drivers or women who cooked food. These relationships and the resultant work opportunities formed a large part of children's livelihood strategies and promised provision of money and food in the short-term. As two young people explained:

RA: What binds you and Mama Ntilie[10] or vegetable sellers?
Daudi[11]: With Mama Ntilie and mother[12] who sells vegetables; even at the moment now you can collect somewhere a new cloth and sell it to them.
Omary[13]: Or she can tell you to help her carry bananas.
Daudu: Yes or [she may ask] 'can you help me to sell [things for] my business I want to go somewhere to get something'?

> Omary: Or it can be common thing to do, there were others selling there and these mama were asking them 'help me to sell over here' and they [the women] go to do something else.

Since street-living children and young people are heavily stigmatised in Tanzania, I was keen to understand why some of the young people appeared to have good relationships with women in the market-place, and why the women would often trust them enough to guard their food stalls; something that was contradictory to the experiences of Pauli and the other participants in his group interview. Daudi and Omary explain the reciprocal relationship that they had built with women in the market-place; bringing them goods to buy, carrying their produce and watching their stalls while they ran errands. These transactions often compounded to build trust between street-living populations and vendors in the market-place that facilitated further opportunities for low-paid work. Those who claim to have benefitted most from these relationships were those who were willing to alter their behaviour to please their benefactors.[14] One child was warned against spending time with the other street-living children and instead spent most of his time with a group of men who drove motorbike taxis:

> Simon[15]: You may also find a brother[16] telling you stop doing a certain thing it will affect you and it is true that you will be affected.
> RA: Are those brothers the street-living adults?
> Simon: No are just people who pass by like motorbike taxi drivers, over there, I spend time with [them].

Omary explains how his intention was to earn himself a good reputation by visibly working hard at the market. He boasted that this ensured he always had work to do, since people were able to trust him.

> Omary: [I] am already known because I used to carry luggage [loads], and apart from that I was selling fruit. [...] The same with getting goods, since I was well known already going to farms and carrying luggage and so forth [he is trusted to be given a bulk of goods to sell and pay for those goods later]. My intention was to be well known and get rid of this name 'street child'. That was how I got to be well known even today, until tomorrow, in the morning I can wake up and take my push cart [to collect and sell goods].

Simon and Omary differ in age (16 and 23, respectively). Simon is comfortable to spend time with the other adults in the market-place and heed their advice regarding who he should avoid socialising with. He also runs small errands for motorcycle taxi drivers and shop keepers in return for food, protection and money. Such errands may be considered trivial or demeaning to an older youth, but Simon's younger age allows him to benefit from the charity of others and he does not yet aspire to higher-paid work. Omary carries the confidence of older age and has spent time building relationships of trust with others in the market-place by working hard. For Omary, earning himself a good reputation in the market, combined with his entrepreneurialism, has enhanced his livelihood opportunities and he does not complain about lacking work, unlike many of the other research participants. Simon and Omary conform to Tanzanian cultural values and expectations that require children and young people to be obedient, hard-working and respectful towards their elders. The relationships of reciprocity that Simon and Omary have developed on the street provide many benefits to their survival and can help them realise their entrepreneurial aspirations. However, the contract of reciprocity that underpins these relationships most likely functions in the short-term, and low-paid temporary work is less dependable than 'employment'[17] which Pauli is seeking (see the previous section). In the relationships that Simon and Omary have built with other adults in the market-place, there is no sense of investment in the young people's futures (e.g. by paying for education or training). It would not be of benefit to the vendors in the market-place to make such an investment since a street-living child is not likely to support their benefactors in old age nor receive inheritance from them, as such arrangements are difficult to regulate outside of family and community networks of accountability (Lassen and Lilleør, 2008). This suggests that there are limits to which street-living children's relationships with members of the community can replace the longitudinal commitment of the intergenerational contract. Street-living children whose ties with their family have been loosened during their time on the street, but who do not fall under the care of an NGO or state service, find themselves in a web of short-term reciprocal relationships that all require maintenance. Consequently, they possess only a fragile network of support which may not withstand significant shocks that inhibit the child's reciprocity, such as illness of a child, a child being imprisoned, or a child's tarnished reputation caused by engagement in theft or other 'antisocial' behaviour.

Who is responsible for street children?

According to the UN Convention on the Rights of the Child and the ACRWC it is families and governments who are responsible for the welfare of children. Many in Tanzania would also agree that this is the case. However, the failure of child rights discourses to bring assistance to street-living children in Tanzania is in part due to the way local services and individuals negotiate their understanding of relationships of reciprocal responsibility and the notion that children have certain inalienable rights. In societies where assistance is assumed to be reciprocal, helping others based on arguments of entitlement may be jarring. It is arguable that accepting child rights from a viewpoint of intergenerational reciprocity makes children's rights conditional and therefore of little value. However, the implicit foundational philosophy of the UNCRC and the intergenerational contract may be similar: while a Western concept of children's rights does not require children to repay what is considered their entitlements (such as food, shelter and protection), one could argue that a neoliberal agenda seeks to nurture children who will benefit nations' economies in the future. This agenda may also be mapped onto NGO programming which is concerned with value-for-money and investing in those who show the most promise of 'progress' (Beazley, 2003), requiring that the most vulnerable either conform to existing provision or become marginalised. This position of marginalisation is where many street-living children and young people find themselves: too difficult (or expensive) to assist by Western neoliberal standards and simultaneously outside of Tanzanian familial support structures of reciprocity, thereby left to take responsibility for themselves. This, perhaps unintentional, step into apparent independence requires children to develop numerous interdependent relationships on the street in order to survive. Children and young people may find that they break familial ties, therefore absolving themselves of responsibility to provide for their family in old age, but consequently they do not have a benefactor who can invest in them and their futures. However, it is doubtful that children's decisions to leave home are due to an explicit defiance of their responsibilities. In many cases, children may consider that it is parents who have neglected their responsibilities to invest in their children, as the following participant explains:

James[18]: If I go back and mother tells me maybe she can bring me to school again, I will go. But if it is just performing chores at home; no.

James desires his mother to invest in his future by sending him to school. If his mother would be willing to send him to school, he would consider returning home to live with her. He does not, however, want to spend his time *just* doing chores to support his family. Some participants experienced a more explicit rupture in the intergenerational contract:

Frank[19]: But there is this, in that situation if there is problem between you with people at home.
Don[20]: Or is like that, father told you to leave.
Kasim: You will no longer stand at his side anymore.
Frank: Yes, you can't listen to your relative anymore.

As Frank, Don and Kasim explain, an explicit rejection from a parent or family member may lead a child to cut ties with this person and no longer consider their opinion in life decisions. Twum-Danso (2009) also argues, in the case of girls in Ghana, that when parents fail to meet children's needs they look to older 'boyfriends' on the street to care for them, thereby switching their loyalty.

There is no easy replacement for the long-term commitment a child or young person should expect from their family, so the breakdown of the intergenerational contract can cause irreparable damage to a child or young person's livelihood prospects (van Blerk & Ansell, 2007). In this respect, preventative measures are required that assist families before children leave home. Such assistance may come in the form of emotional or financial support for parents, mediation during family disagreements or protection of children's rights to inheritance following the death of a parent. More research is required regarding the reasons for, and means of protection against, families' inability to uphold their responsibilities towards their children and dissolution of the intergenerational contract.

NGOs also have a role to play in mediating relationships between street-living children and their families as well as the provision of care when families are unable or unwilling to provide this support. One intervention in Puebla, Mexico, by an organisation called JUCONI, invites street-living boys to live in an open-door community house, and coordinates frequent visits to the boys' families (Schrader-McMillan and Herrera, 2016). The premise of the intervention is to work alongside, and support, the families of street-living children since JUCONI believe that the parents' unmet needs mean that they are not able to provide for their children; therefore, they work through a process of 'reparenting the parent' (Schrader-McMillan and Herrera, 2016: 255). JUCONI programmes aim to reintegrate children into their family home but where this is not possible provide young people access to a 'youth house' until they are able to be self-sufficient; thus, taking responsibility for children and young people while respecting their agency and autonomy to choose when to become independent. Hecht (1998), from his research with street-living children in Brazil, noted that the organisations that appeared to have most success with regard to keeping children away from the street for extended periods of time were residential in nature and appeared to have no limit for children and young people's length of stay, suggesting that these interventions provided children with a level of stability. Hecht expresses doubt, however, that these services would suit many street-living children he encountered during his research due to their requirement of abstinence from drug use, which is difficult for many children and young people to adhere to. A contrasting example of NGO assistance for street-living populations involves providing supported living arrangements for alcohol-dependent adults in Canada (Evans et al., 2015). Although this intervention is targeted at adults in the Western world, it suggests a means of supporting homeless populations with addictions that respects a person's right to shelter and care as well as, in some places, a dignified environment in which to die. Such assistance appears relevant to young people involved in my study in Tanzania, some of whom warned that they risked becoming dependent on drugs and 'giving up on life'.

It seems that families, governments and civil society share legal and moral responsibilities for the welfare of children and that issues arise when these systems of support are not sufficient to meet all children's needs. Sustained and dependable support is particularly important to children so they may invest in their own skills, broaden their network of opportunities and facilitate their move into independent, self-sufficient adulthood. Tanzanian street-living children and young people's challenges in receiving the support they need to transition them into self-sufficient adulthood mirrors the experiences of other youth across Africa (Langevang, 2008; Day and Evans, 2015). This suggests that the root causes of poor prospects among street-living children and young people may be due to systemic issues relating to the disruption of family support structures, neoliberal reform and reduced investment in children's skills, education and entrepreneurism through economic hardship, rather than issues of individual culpability.

However, it is important to understand that street-living children and young people are not merely helpless victims. As Omary demonstrated, it is possible for children and young people to use their agency and ingenuity, perhaps leveraged by good fortune (Mizen, 2018), to work their way out of the street and poverty by harnessing a web of relationships with business owners in commercial areas. However, the stigmatisation that street-living children face owing to their presence on the street and apparent estrangement from their families places them at a considerable disadvantage when seeking access to secure employment and training. Such limited opportunities lead many street-living children and young people to engage in risky income-generating opportunities that lead to further stigmatisation and marginalisation of this population.

Conclusion

Street-living children and young people are a marginalised population, many of whom find themselves entirely responsible for meeting their own daily needs. Street-living children build relationships with peers on the street and adults in the market-place and form relationships of reciprocity which help them to access the food, shelter and income that they need for survival. However, this population struggles to find relationships that can replace the role that their family would have played in providing for their futures. Alongside the practical challenges of taking responsibility for day-to-day needs, street-living children and young people are heavily stigmatised in Tanzania due to their appearance, association with theft and drug use, and, most notable for the discussion in this chapter, the perception that they have abdicated familial responsibility. This stigmatisation exacerbates their exclusion from society and decreases the amount of empathy, and therefore assistance, they may receive from the communities they encounter.

While children would ordinarily be the responsibility of their families under cultural traditions of intergenerational reciprocity, and the responsibility of the government as stipulated in the UNCRC, both systems have failed to care for the needs of street-living children and young people in Tanzania. NGOs have a crucial role to play in extending the governmental mandate to protect and provide for children and young people; however, programmes require funding and resources to replicate the role of the parent in caring for a child's current needs as well as their training and employment prospects. While it is important to provide alternative care arrangements for those who are unable to live with their families, it is prudent to consider why parents struggle to ascertain lines of responsibility for children when a marriage, or parenting partnership, disbands. Children and young people living on the street shoulder the additional burden of stigmatisation that results in their further social marginalisation, in part exacerbated by the blame attributed to them: blame that is used to dehumanise those living on the street and absolve the wider community of their responsibility to care for them (Belcher and DeForge, 2012). While interventions are needed to help assist families in supporting their children and to provide for children and young people who are no longer supported by their families, children's responsibilities to their families and local perceptions of blame require additional deconstruction if the root causes of the stigmatisation of street-living children and young people are to be challenged.

A note of acknowledgement

I would like to thank the children and young people who contributed to the research included in this chapter, as well as my research assistants, Asimwe Suedi, Fred Mbise and Respick, who performed the roles of translator, transcriber and gatekeeper. Thank you also to Nadia von Benzon and Catherine Wilkinson who provided comments on earlier drafts of this chapter, as well as the participants in the 'Other Childhoods' session in the RGS-IBG Annual Conference 2017, who offered input into the initial conceptual ideas that provided the foundation for the finished piece.

Notes

1 Various terminology is used in literature to define street children, including street-connected, street-involved and children in street situations. The children and young people engaged in this research had migrated to urban areas and lived and slept on the street full-time, with the exception of those who had earned enough money to rent accommodation with peers. In this chapter, I will refer to the participants as street-living to distinguish them from children and young people who work on the street but return to an immediate, or wider, family home in the evening.
2 The location of the research is omitted for reasons of confidentiality.
3 Pseudonym.
4 Aged 17, street-living.
5 Research assistant.
6 Aged 14, street-living.
7 Male, age 19, renting a room.
8 Male, age 22, interview 7, renting a room.
9 'Mother' is a term used to address all older women and does not denote a biological relationship.
10 These are women who sell cooked food in the market, literal translation: 'Mama fill me up'.

11 Aged 19, street-living young person.
12 Non-biological relationship with older woman.
13 Aged 23, former street-living young person.
14 See too the story of Ama, a Ghanaian unaccompanied child migrant in Boampong's chapter, this volume.
15 Pseudonym. Male, age 16, street-living.
16 'Brother' is a colloquial term for a close friend, or respected peer.
17 The young people from Pauli's group interview desire to be employed in a trade (e.g. as a mechanic) or as a bus conductor. Conversely, the work that Simon and Omary engage in is independent and entrepreneurial.
18 Pseudonym, male, age 16, interview 14, renting a room.
19 Pseudonym, male, age 23, interview 13, renting a room.
20 Pseudonym, male, age 16, interview 13, street-living.

References

African Commission on Human and Peoples' Rights (1981) *African Charter on Human and People's Rights*. Available: www.humanrights.se/wp-content/uploads/2012/01/African-Charter-on-Human-and-Peoples-Rights.pdf (accessed 8 June 2018).

Awedoba, A.K. (2002) *Culture and Development in Africa with Special References to Ghana*. Accra: Institute of African Studies, University of Ghana Legon.

Beazley, H. (2003) 'The construction and protection of individual and collective identities by street children and youth in Indonesia.' *Children Youth and Environments* 13(1): 105–133.

Belcher, J.R. and DeForge, B.R. (2012) 'Social stigma and homelessness: The limits of social change.' *Journal of Human Behavior in the Social Environment* 22(8): 929–946.

Blackford, J., Di Cecio, L., Jerving, N., Kim, S. and Sayagh, S. (2008) *Supporting the Transition of Street Children to Self-Reliant Adulthood in Tanzania*. Available: https://sipa.columbia.edu/file/3226/download?token=Sjyl7IVi (accessed 21 June 2018).

Concicini, A. (2008) 'Surfing in the air: A grounded theory of the dynamics of street life and its policy implications.' *Journal of International Development* 20(4): 413–436.

Davies, M. (2008) 'A childish culture? Shared understandings, agency and intervention: An anthropological study of street children in northwest Kenya.' *Childhood* 15(3): 309–330.

Day, C. and Evans, R. (2015) 'Caring responsibilities, change and transitions in young people's family lives in Zambia.' *Journal of Comparative Family Studies* 46(1): 137–152.

Evans, R. and Becker, S. (2009) *Children Caring for Parents with HIV and AIDS: Global Issues and Policy Responses*. Bristol: The Policy Press.

Evans, J., Semogas, D., Smalley, J.G. and Lohfeld, L. (2015) '"This place has given me a reason to care": Understanding "managed alcohol programs" as enabling places in Canada.' *Health & Place* 33: 118–124.

Frankenberg, S.J., Holmqvist, R. and Rubenson, B. (2010) 'The care of corporal punishment: Conceptions of early childhood discipline strategies among parents and grandparents in a poor and urban area in Tanzania.' *Childhood* 17(4): 455–469.

Glaser, B.G. and Strauss, A.L. (1967) *The Discovery of Grounded Theory: Strategies for Qualitative Research*. Chicago: Transaction Publishers.

Hecht, T. (1998) *At Home in the Street: Street Children of Northeast Brazil*. Cambridge: Cambridge University Press.

Hecker, T., Hermenau, K., Isele, D. and Elbert, T. (2014) 'Corporal punishment and children's externalizing problems: A cross-sectional study of Tanzanian primary school aged children.' *Child Abuse & Neglect* 38(5): 884–892.

Hollos, M. (2002) 'The cultural construction of childhood changing conceptions among the Pare of Northern Tanzania.' *Childhood* 9(2): 167–189.

Langevang, T. (2008) '"We are managing!" Uncertain paths to respectable adulthoods in Accra, Ghana.' *Geoforum* 39(6): 2039–2047.

Lassen, D. and Lilleør, H. (2008) *Informal Institutions and Intergenerational Contracts: Evidence from Schooling and Remittances in Rural Tanzania.* University of Copenhagen. Available: www.econ.ku.dk/cam/wp0910/wp0708/2008-03.pdf/ (Accessed 14 June 2018).

Libório, R.M.C. and Ungar, M. (2010) 'Children's labour as a risky pathways to resilience: Children's growth in contexts of poor resources.' *Psicologia: Reflexão e Crítica* 23(2): 232–242.

Mizen, P. (2018) 'Bringing the street back in: Considering strategy, contingency and relative good fortune in street children's access to paid work in Accra.' *The Sociological Review* 66(5): 1058–1073.

Organisation of African Unity (1990) *African Charter on the Rights and Welfare of the Child.* Available: www.unicef.org/esaro/African_Charter_articles_in_full.pdf (accessed 14 June 2018).

Schrader-McMillan, A. and Herrera, E. (2016) 'The successful family reintegration of street-connected children: Application of attachment and trauma theory.' *Journal of Children's Services* 11(3): 217–232.

Shand, W., van Blerk, L. and Hunter, J. (2017) 'Economic practices of African street youth: The Democratic Republic of Congo, Ghana, and Zimbabwe', in: Abebe, T., Waters, J. and Skelton, T. (Eds.), *Labouring and Learning, Geographies of Children and Young People.* Singapore: Springer, pp. 411–431.

Twum-Danso, A. (2009) 'Reciprocity, respect and responsibility: The 3Rs underlying parent-child relationships in Ghana and the implications for children's rights.' *The International Journal of Children's Rights* 17(3): 415–432.

van Blerk, L. (2012) 'Berg-en-See street boys: Merging street and family relations in Cape Town, South Africa.' *Children's Geographies* 10(3): 321–336.

van Blerk, L. and Ansell, N. (2007) 'Alternative care giving in the context of AIDS in southern Africa: Complex strategies for care.' *Journal of International Development* 19(7): 865–884.

Wells, K. (2015) *Childhood in a Global Perspective.* Second edition. Cambridge: Polity Press.

4 Subverting neighbourhood normalcy and the impacts on child wellbeing in Malta

Bernadine Satariano

Chapter summary

Children's health and wellbeing are influenced by multiple factors in different environments including the family, the neighbourhood and the wider social context. This study, through in-depth interviews with children belonging to different familial compositions, explores how within three deprived Mediterranean Maltese neighbourhood communities, children's wellbeing is affected in diverse ways. It explores how the marital status of the parents is perceived in the neighbourhood's cultural and social context and how the child experiences repercussions of the familial status, which indirectly determines the child's wellbeing. It emphasises how the history of the place shapes the norms which are socially accepted in the locality, thus highlighting that health and wellbeing is contingent to place. Moreover, the socio-economical changes that occur across time in a place may alter the norms and affect people's wellbeing in different ways. It also shows how children are affected by the prevalent gendered norms and processes transmitted in the neighbourhood and how they are contingent upon the neighbourhood in which one lives. This emphasises that local place has an important role on the health and wellbeing of children, even in relation to their familial marital status.

Introduction

Effects of family status on child wellbeing

In recent decades, developed countries have experienced dramatic transformations in family structures leading to a change in the definition of the family. It is becoming more common that children experience various family living arrangements over their life course, even if born to married parents (Haskey, 1997). Burgess (1948) saw changes in family types as positive, evolving and adapting as part of wider social change. Yet, several debates see the changes in the family structure as negative (Murray et al., 1994; Fevre, 2000). According to Bernardes and Smith (1993), the traditional, nuclear family is perceived as the most powerful, while other family structures are considered problematic.

Numerous studies analyse how changes in family structure impact on child wellbeing. There is an emphasis that traditional families with married parents have the best familial environments for children (Amato, 2005; Kelly, 2000; Cummings and Davies, 2011), while children living in cohabitant and single parent families tend to fare worse (Sigle-Rushton and McLanahan 2004; McLanahan and Sandefur, 1994). Research also suggests that children living in single parent and cohabitant families are more likely to experience multiple transitions and are thus more prone to family instability (Craigie, 2008). Family transitions also impact on the cognitive and behavioural effects of children, as they are subject to disruptive behaviour at school, poor emotional adjustment and lower educational achievements (Martinez and Forgatch, 2002). Although the number of single parent families are increasing, in general they undergo many challenges, including: financial difficulties, limited social support, social stigma and limited time spent with their children (Coles, 2015; McLanahan and Sandefur, 1994).

Another body of research analyses the effects of parental conflict on child wellbeing as a risk factor, emphasising that when there is frequent and intense conflict between parents, children and adolescents are likely to display internalising problems (e.g. depression and anxiety) and externalising problems (e.g. aggressive behaviour and criminal activity) (Buehler et al., 1997; Dadds and Powell, 1991; Di Manno et al., 2018; Emery, 1982; Prince, 2018). Marital conflicts are likely to increase emotional and behavioural problems (Fishman and Meyers, 2000), affect children's emotional security, and impact on their academic achievement (Timmons and Margolin, 2014), thus children may take different time frames to adjust.

The neighbourhood social context and its impact on child wellbeing

Studies rarely assess how family structures impact on child wellbeing within a neighbourhood context. The area of residence may have significant influences on health and wellbeing, over and above the effects of individual characteristics (Macintyre et al., 1993; Jones and Moon, 1993; Curtis, 2010; Gatrell and Elliot, 2015). Differences in population health between places could be because of differences in the people who live there (compositional explanation) or because of differences between the places (a contextual explanation). There are contextual processes operating at the scale of whole communities or geographical areas which are important for health and health inequalities (Cummins et al., 2007; Macintyre et al., 1993). Studies have demonstrated the importance of contextual effects in relation to a range of health outcomes, including self-reported health (Cummins et al., 2005), mortality (e.g. Waitzman and Smith, 1998) and morbidity (e.g. Shouls et al., 1996), and health behaviours, such as smoking (e.g. Duncan et al., 1999). Macintyre (1997) discusses how 'collective' factors can contribute to explanations of observed geographical variations in health outcomes. Attending to collective factors means examining the significance of shared norms, traditions and values, and recognising the importance of those socio-cultural and historical features that may be shared within a community (Macintyre et al., 2002). Including collective factors in an analysis of geographical variation of health outcomes and behaviours supports the argument that context and composition should be engaged through a multi-faceted approach. This emphasises that places can be active contributors to the social processes occurring within that same place and places may also constrain social processes or produce and reinforce social structures (Curtis, 2004).

Research has shown that there are important contexts within which children and adolescents are socialised into certain values, opinions and attitudes in their everyday lives. These include the physical contexts (such as residential, public spaces), institutional contexts (youth clubs, sports clubs, and school), as well as social contexts (family, peers, teachers, and friendly adults in the neighbourhood) (Abbott-Chapman and Robertson, 1999; Duncan and Raudenbush, 1999; Leventhal and Brooks-Gunn, 2000; Bowen et al., 2002; Irwin, 2009; Wen et al., 2009; Theron et al., 2011). The research discussed in this chapter analyses how children and adolescents' social context, in relation to the marital norms, impacts on their health and wellbeing. Indeed, children too are affected by the local neighbourhood social norms. The local transmissions of social norms through childhood socialisation constitute an important part of how neighbourhood characteristics and processes of daily life are reproduced.

The health and wellbeing of children and adolescents is influenced by neighbourhood norms, procedures and rules (Baron, 2004). Children, apart from acquiring the norms of their parents, are also influenced by their peers in the immediate social environment, while they may be exposed to their neighbourhood environments through schooling or other activities which may transmit diverse social norms. For example, Leventhal and Brooks-Gunn (2000), Katz et al. (2001) and Galster (2012) highlight that children living in deprived neighbourhoods are prone to being exposed to antisocial behaviour amongst peers which may impact on their wellbeing (Galster, 2012; Lorenc et al., 2012), their academic performance (Lord and Mahoney, 2007; Galster, 2012) and verbal ability (Sampson et al., 2008). If young people notice that their peers engage in certain health risk behaviours, there is a higher probability that they will engage in this behaviour, as it is perceived as normal (Olds and Thombs, 2001; Rimal and Real, 2003; Baker et al., 2003). On the other hand, socialising with a peer group with positive health behaviours may develop norms of positive health behaviours. Some social neighbourhood norms operate for the benefit of neighbourhood inhabitants, such as the prevention of deviant behaviour (Horne, 2004) or the formation of socially cohesive communities (Hechter and Opp, 2004). Social support amongst peers can be linked to positive wellbeing, by providing buffering effects against difficult circumstances (Rigby, 2000). However, some social norms of social cohesion may harm the health and wellbeing of individuals, including children, through features of shame, social exclusion and punishment (Foley and Edwards, 1998, Leavitt and Saegert, 1990; Hammond and Axelrod, 2006).

This chapter will emphasise that children are affected by neighbourhood social norms in relation to family structures (see also Pearson's chapter on intergenerational reciprocity and street children, this volume). It seeks to increase understanding that the geographical social context is important in determining the health and wellbeing of children. In this case, the marital status of the child's family is valued differently across different geographical contexts. Therefore, children's health and wellbeing are not only determined by parental marital status, but also by how the parental status is perceived within the social norms of the neighbourhood.

The social change in family life within a Maltese southern European context

The role of the family is important for an individual's health and wellbeing in most societies, yet across different cultures the idea of the family varies. Within a Maltese culture, often one's identity is defined by the type of family that person belongs to. Therefore, the honour of the family is important (Abela et al., 2005; Satariano and Curtis, 2018).

Malta is an interesting case study because it has a traditional, Mediterranean, Roman Catholic culture, with close knit family ties and social relations. Since Malta is the smallest and most densely populated country in the European Union, the levels of social interaction and social cohesion are high (Tabone, 1995; Satariano and Curtis, 2018). Due to its traditional values, family as an institution plays an important role in society (Abela et al., 2005). Moreover, the extended family has an important influence on the nuclear family (Satariano and Curtis, 2018).

Recently the rate of social change in attitudes to marriage in Malta has accelerated. Divorce legislation came into effect in 2011. Before this, Malta experienced an increase in numbers of unmarried mothers (Abela et al., 2005). Moreover, in 2014 the civil union bill was passed, leading to the legalisation of same-sex marriage in 2017. These legislative changes affecting the institution of the family are being debated in the public domain at a national and local level, with questions over what *family* means.

As family life in Malta is experiencing a shift reflected by these legislative measures, this study seeks to explore life experiences within an environment of social change, and analyses the dynamic processes occurring within deprived neighbourhoods in Malta and how they impact on the health and wellbeing of children. This is because the local place may operate according to specific social norms which may be diverse across places. Therefore, assessing the child's wellbeing within different familial structures should also consider the role of local social processes and norms.

Methodology

This study aims to reveal the nuanced experiences of children according to their familial status and neighbourhood environment. From a social geographical aspect, it tries to interpret life experiences within the wider social and spatial setting (Jackson and Smith, 1984; Smith, 1984; Pile, 1991). Through in-depth interviews, this study explores the lived experiences of children, aged 5–16, belonging to different family structures living in three deprived neighbourhoods in Malta. Fifty-four children and adolescents, male and female, belonging to 30 families, participated in this study. A purposive sampling approach was used, recruiting children belonging to married, single parent and cohabitant families from each locality. This approach aimed to capture the diversity of family life during social change within their neighbourhood environments.

Little research has been conducted on children's wellbeing within a neighbourhood context in Malta. The research areas chosen for this study are considered in need as a local government agency is located in these neighbourhoods and is responsible for social welfare services, employment opportunities, youth centres and adult training. This government agency was used to recruit the target sample children. To protect the identity of respondents, the names of the neighbourhood and children are anonymised. The research areas chosen for this study are a 'traditional walled' town; a 'deindustrialised' town; and a 'modern' town. Although the localities have a high proportion of families in need of social assistance, the 'traditional walled' town and the 'deindustrialised' town have a long history of settlement, whilst the 'modern' town is a recent residential area.

Ethical approval was granted by the researcher's institution, Durham University. Consent was obtained from parents to interview each child and assent was provided by the children themselves. Ice breaker activities (drawings, mind maps and story games) together with props (soft toys) were used to create a non-threatening environment and to diminish the power imbalance between adult researcher and child participant (Christensen, 2004; Driessnack, 2006). Parents were interviewed to understand the family context from the parent's viewpoint. The interviews were held in coffee shops or in playgrounds chosen by the children themselves. All interviews were conducted in Maltese and were digitally recorded.

This study makes use of a constant comparative approach. Comparing the three study areas helps to demonstrate the impact of family composition within diverse social neighbourhood norms. This will allow a better understanding of the variability in local area conditions and how norms, in relation to familial marital status, impact on child wellbeing.

Findings: how children's wellbeing may differ according to the neighbourhood social processes and contextual norms in relation to their parents' marital status.

This section will explore how the marital status of the mother is perceived in the neighbourhood and how the children experience the repercussions of the familial marital status, which indirectly determines their health and wellbeing. The discussion will explain how children are affected by the prevalent norms and processes transmitted in the neighbourhood and emphasise that they are contingent upon the neighbourhood in which one lives.

The 'traditional walled' town

A high percentage of elderly and their extended families reside in the 'traditional walled' town. Considering the Maltese Mediterranean context, religious norms still influence the attitude of the elderly relatives and members of the extended family, and in turn impact on the views of mothers with regards to separation, divorce or cohabitation.

Due to the pressures imposed by elderly relatives, children living in the 'traditional walled' town have described how, when their parents separated, they endured a painful experience because they stood out in their community. This is because the extended family and the elderly in the neighbourhood feel that the community should hold strong religious values and norms. To maintain these norms relatives and community members exercise high levels of social control and try to reunite separated couples, since it is perceived as an act against Roman Catholic religious norms and values.

This was illustrated by Isaac when he recounted that his mother separated from his father, due to his father's financial debts, and started dating a separated man from the same town. Despite the husband's irresponsible behaviour, community members felt that it was Maria, Isaac's mother, who was crossing the boundaries of accepted social norms and that she was at fault. '*I know that my mother has been through a lot because of my father's financial problems and sometimes I myself told her to go out with friends as she was feeling down*' (Isaac). Although at first Maria did not give heed to how the people in the neighbourhood judged her, it was when her sons started being bullied by other teenage boys because of their mother's extra marital affair that she ended her new relationship. Isaac explains. '*Things with our friends were going really bad as she was dating the father of two boys from Valletta. Everyone turned against me and my brother*'. Maria realised that going against the social conventions of marital norms provoked these bullying tactics towards her and her children. She realised that she had to stop her relationship when the community was putting pressure on her and her children, consequently harming their wellbeing. This narrative sheds light on how the neighbourhood manipulates the emotions of Isaac's mother who was ready to sacrifice her interests in favour of her children in relation to marital norms.

For Isaac, the fact that his mother stopped her extra marital relationship was not enough for him to feel part of the community, abiding by the norms. He states, '*Now it's better as they are not dating any longer but I still wish that we were like a normal family with our father living with us. I am trying to convince my father to behave better so that my mother would accept him back*'. (Isaac)

Due to the social norms, the neighbourhood imposes responsibility on the child to try to mediate so that the parents get back together after separation.

Isaac narrated how his aunts and uncles were trying to encourage him to convince his mother to accept his father back. They also gave him tips on how to persuade his father to change his behaviour so that he could return to live with them as a united family. In the second round of interviews Isaac stated that, '*Now I feel great!*' as his parents are back together. Isaac feels accepted back in the neighbourhood, he is glad that his efforts to reunite his parents were successful. This shows that the neighbourhood practices such powerful social control that it does not only exclude the mother due to her extra marital affair, but also puts pressure on children, coercing them to try to reverse their mother's actions. This indicates that the community has dominant features of social control to retain the social neighbourhood norms.

Maya, another adolescent from the 'traditional walled' town, feels inferior to other children in the neighbourhood because her parents separated and both have started new relationships. She commented, '*It was so embarrassing on my Holy Confirmation; having my mother with other children and her boyfriend and my father with his girlfriend and their baby sitting on the same bench in the church! Even though they behaved really well and they did not argue or fight or anything, it feels humiliating not having only your mother and your father. We were so different from the rest and that made me feel really uneasy.*' (Maya)

Although Maya showed understanding towards her parents' decision to separate, she experiences criticism by the neighbours who make her stand out in the community when compared to other children in the neighbourhood. Maya explains how she and her grandma noted that people in their parish noticed that Maya's parents started new families, leaving her with her grandma. They passed comments of pity to Maya, implying that she had been abandoned. This means that children in similar familial situations may be highly affected by the social control exerted by the socio-religious norms in their neighbourhood. These socially controlling, punitive norms are harmful towards child wellbeing, suggesting that children are at risk because their parents are not abiding by the neighbourhood's familial norms. The levels of social control and the importance of adherence to religious, marital norms are so strong in the neighbourhood that they are transmitted in a harmful manner to children and adolescents.

The 'deindustrialised' town

The narratives of the children living in the 'traditional walled' town differ from those in the 'deindustrialised' town. As highlighted earlier, the 'deindustrialised' town experienced massive unemployment where the male breadwinner stopped bringing income home and started being perceived as a worthless member of the family. The mothers realised that if they separated and adopted single mother status they would be financially better-off, since they would start receiving single mother benefits (Satariano, 2016). Therefore, the deindustrialisation of this locality altered the social norms in relation to marital status and the meaning of the family that provides for children. Most of the children in the deindustrialised town appreciate that their mother separated from their father and adopted a single mother status. They are aware that their mother tries hard to provide for them in the absence of their father.

> *I really appreciate what my mother does for us, because she goes to great pains to take care of us single-handedly. I feel that now that Mum left Dad we are living a better life.* (Charmaine)
>
> *My father used to return home drunk every day, because he had nothing else to do! It used to be scary, he was always fighting. He never worked and never earned money. Now he does not live with us and that's good. Life is more calm like this.* (Gillmor)

Therefore, within this 'deindustrialised' town, single parenthood is considered the ideal family status, transmitting positive effects on health and wellbeing.

The following accounts help to explain why children may feel relieved that their father has left their household, since they were experiencing feelings of abuse. The children are aware that most men in the neighbourhood behave in an abusive manner. Backed by their extended family, the children in this neighbourhood are preventing their mother from starting a new relationship, thereby safeguarding their own health and wellbeing. They are content to live without the presence of a male figure, free of conflict, violence and victimisation. Indeed, Nathan, Brian, Kevin and Gillmor stated that if their mother were to start a new relationship, even if not cohabiting with him, they would feel betrayed, for her attention would be directed towards her new partner and not towards them. In their opinion this would have an ill effect on their wellbeing:

> *I'm sure that if Mum had to find another partner, only God knows what would happen to us... how we'll end up! But Mum prefers to look after us.* (Kevin)
>
> *There are mothers who find it difficult to bring up their children on their own, so they seek support by finding another partner. Yet the children of these families are unhappy because the new partner, not being their father, does not love them.* (Nathan)

The assumptions that the mother's new partner is often aggressive, violent and abusive are based on observations of other children in the neighbourhood who withstand physical abuse and aggression from their mother's partner:

> *I saw one of my classmates who is usually a bully, crying and sobbing alone in the street, because his mother's boyfriend had given him a sound beating.* (Roberto)

In this 'deindustrialised' town, when a mother starts a new relationship, the extended family tends to care for the children, due to fears of domestic violence or abuse of the partner towards the children. To put pressure on the mother to leave her cohabitant partner, the extended family instigates anger in the children towards their mother, making them feel betrayed by her. Amanda narrates that, since her mother started dating a new partner, she and her brothers are treated inferiorly to her mother's partner.

> *Mum practically abandoned us. After dad left her she found this boyfriend who is an alcoholic. She turned into an alcoholic herself and she cannot even take care of us. My brothers are all the time fighting between themselves and with my mother. My grandma and my aunt take care of us and they tell us she prefers him and not us.* (Amanda)

What is clear is that in this 'deindustrialised' town, the mother's role encompasses every need of the child, while little is expected from the father. The culture of the neighbourhood has fostered these gendered roles to the extent that when the father leaves the family, the children are hardly emotionally affected, but if the mother decides to start another relationship, even if she does not leave the household, the children feel unloved and abandoned.

The 'modern' town

In the 'modern' town, being multi-ethnic and touristic, there are low levels of social cohesion. It emerged that there are few socially controlling norms experienced and transmitted in the locality. The Roman Catholic norms are not as strong in this neighbourhood and the experiences of social pressures and controls from the extended family narrated by the children in the other neighbourhoods were not pointed out. For this reason, children are not experiencing judgmental attitudes from the neighbours.

However, this lack of social control can have diverse effects on children's wellbeing. The combination of these neighbourhood processes is not helping fathers to comply with Maltese norms, especially that of providing for their children. On the contrary, these processes absolve them from their responsibility as fathers. Within this context, the father may disappear from the life of his children without being condemned by the community. This leaves single mothers in this 'modern' town to cope on their own with the upbringing and welfare of their children. Moreover, these families, apart from having no support from the father, also do not receive support from the neighbourhood, due to lack of social cohesion, nor from their extended family, as it is at a physical and social distance from the nuclear family. These families have limited interactions with their extended family, unlike those of the 'traditional walled' and 'deindustrialised' towns where the extended family is likely to support the children.

Several children from this 'modern' town complained that, due to the absence of a supportive father, their family's financial situation is poor and they are deprived of both the necessities and luxuries other children in their neighbourhood enjoy. Their wellbeing is thus tarnished due to their experiences of relative inequality.

Norma's daughter explains that she feels inferior to her friends as she cannot have a birthday party like them, nor can she attend other children's parties as her mother cannot afford to buy presents for the party hosts. She also feels that she does not have proper clothes to wear for parties like other children do. She would have liked for her father to be present as she feels that this would have diminished her hardships.

> *I know girls in my class whose father does not live with them, so that does not bother me… but I wish that my father would pay and throw me a party and I can invite friends like they do. Not all fathers are like my father, some do remember their children's birthday and still give them presents… mine, he does not want to remember me and that hurts!* (Hayley)

Leon and his brothers explained how sad they feel about the neglectful behaviour of their father. They recount, '*We know that our father does not love us, because he knows that we would like to play certain computer games which he has. He never lets us play with them or lends them to us to play at home. That hurts because other children's parents buy them these games*'. This behaviour of neglectful fathers may initiate more serious situations analysed in other research in the United States where it was found that lack of informal social control may result in more cases of child neglect and maltreatment (Guterman et al., 2009; Molnar et al., 2003; Yonas et al., 2010; Emery et al., 2015).

Children living in neighbourhoods with heterogeneous types of families tended to emphasise their feelings of inequality. It emerged that the children most likely to suffer from a sense of relative deprivation are those coming from single parent families. As shown in the quotations above, children of single parents drew attention to the fact that when the father is not contributing to their material needs they feel that the father does not love them. When parents buy things for their children, they want to convey a message that they are providing for their children, and they do not want their children to be inferior to other children, thus transmitting feelings of pride in their children. On the other hand, when the father does not contribute to the welfare of his children, children feel that they are unloved and neglected. The fathers in a separated family are expected to express their sense of attachment to the children by giving them gifts.

Yet, living in an anonymous neighbourhood, lacking social controlling norms, can exert positive effects on the children. Mariah and her siblings, in contrast to the narratives of children in the 'deindustrialised' town, narrated how loving, caring and generous their mother's new partner is. They noted that he is more attentive towards their needs and a much better father than their biological one, and expressed relief that their biological father no longer lives with them. '*He buys us food, clothes, pencils, copybooks for school and when we go out he buys us sweets!*' (Mariah) This means that in a hetero-geneous neighbourhood with low levels of social control there are no judge-mental attitudes towards extra-marital relationships, so cohabitant partners are not inhibited to step outside the conventional norms of wider Maltese society and show affection to their partner's children. On the other hand, children like Hayley and Leon are suffering from this lack of social control in the community, since their father is not feeling admonished by the community for his neglectful behaviour towards them. This means that children living in a neighbourhood with low levels of social control and a lack of dominant norms may affect children both positively and negatively, depending on the personality of the parent or partner.

Conclusion

This chapter argues that to analyse the impact of family structure on child wellbeing, there is a need to focus on the geographical context, taking into consideration the local economic, social and cultural norms. It emphasises that the experiences of children living in different family structures may be experienced differently in diverse neighbourhood environments. Therefore, one cannot assume that living in a cohabitant, single parent or married family has the same effect on all children no matter where they live. The cultural and social norms experienced in a place can have different impacts on children living in similar family structures.

Within a Maltese context, religious norms related to accepted marital status are not all evolving in the same manner across different neighbourhoods. In the 'traditional walled' neighbourhood, the Roman Catholic religious norms are strongly present and through social control parents are encouraged to remain in a married union, despite the difficulties that the couple may be experiencing. Indeed, emerging from the children's narratives, the neighbourhood impacts on their wellbeing as they are being used to retain the religious norms. It is evident in these narratives that when parents go against Roman Catholic religious norms, children experience feelings of stigma.

However, it may be interpreted that the social control that the extended family and the elderly in the neighbourhood are exerting on the couple may be for the benefit of the children. This is because they fear that the children belonging to these families may be abandoned and neglected. Therefore, they use socially controlling norms to encourage the parents to prioritise their par-ental role over their personal needs and invest in better care of and attention towards their children. On the other hand, in the 'modern' neighbourhood, these religious norms are not experienced by the community. This is impacting on children's health and wellbeing, both in a positive and negative manner.

In the 'deindustrialised' town, the interviewed children living in single parent families describe their experience of their family status in a more positive way than do the children living in single parent families in the other two case study towns. The children in the 'traditional walled' and 'modern' towns experience feelings of inequality and inferiority when they compare themselves to other children living in united families. Yet, those children living in the 'deindustrialised' town feel that they fare better when they are living without a man in the household as they are living more serenely. Therefore, children living in single parent families can feel that they are faring the best in comparison to children living in families of other marital status in their neighbourhood.

This chapter has shown that there are differences between the children's attitudes towards their parents' marital status according to their residential neighbourhood environments. If these observations represent the general situation in the study areas, then across different neighbourhood environments, the ideal marital status is transmitted differently to children. A neighbourhood's social norms, shaped throughout its history due to the context and composition of the place, have a significant impact directly and indirectly on the child's wellbeing. Therefore, one cannot assume that children living within the same familial structure have similar experiences with similar impacts on their health and wellbeing. Rather, the influence of neighbourhoods on individuals is dynamic and contingent.

References

Abbott-Chapman, J. and Robertson, M. (1999). Home as a private space: Some adolescent constructs. *Journal of Youth Studies*, 2, pp.23–43.

Abela, A., Frosh, S. and Dowling, E. (2005). Uncovering beliefs embedded in the culture and its implications for practice: The case of Maltese married couples. *Journal of Family Therapy*, 27(1), pp. 3–23.

Amato, P.R. (2005). The impact of family formation change on the cognitive, social, and emotional well-being of the next generation. *Future Child*, 15(2), pp.75–96.

Baker, B.L., McIntyre, L.L., Blacher, J., Crnic, K., Edelbrock, C. and Low, C. (2003). Pre-school children with and without developmental delay: Behaviour problems and parenting stress over time. *Journal of Intellectual Disability Research*, 47, pp.217–230.

Baron, S. (2004). Social capital in British politics and policy making. In: J. Franklin. (ed.) *Politics, Trust and Networks: Social Capital in Critical Perspective*. University of Glasgow: Families and Social Capital ESRC Research Group Working Paper No. 7.

Bernardes, J. and Smith, D.S. (1993). The curious history of theorizing about the history of the Western nuclear family. *Social Science History*, 17(3), pp.325–353.

Bowen, G.L., Richman, J.M. and Bowen, N.K. (2002). The school success profile: A results management approach to assessment and intervention planning. In: Roberts, A.R. and Greene, G.J. (eds.), *Social Workers' Desk Reference*. New York: Oxford University Press; pp. 787–793.

Buehler, C., Anthony, C., Krishnakumar, A., Stone, G., Gerard, J. and Pemberton, S. (1997). Interparental conflict and youth problem behavior: A meta-analysis. *Journal of Child and Family Studies*, 6, pp.233–247.

Burgess, E.W. (1948). The family in a changing society. *The American Journal of Sociology*, 53(2), pp.125–130.

Christensen, P. (2004). Children's participation in ethnographic research: Issues of power and representation. *Children and Society*, 18(2), pp.165–176.

Coles, R. (2015). Single-father families: A review of the literature *Journal of Families Theory and Review*, 7(2). doi: doi:10.1111/jftr.12069

Craigie, T.A. (2008). Effects of parental presence and family instability on child cognitive performance. Working Paper, 2008–2003-FF. Princeton, NJ: Centre for Research on Child Wellbeing.

Cummings, M. and Davies, P. (2011). *Marital Conflict and Children: An Emotional Security Perspective*. New York: Guilford Press.

Cummins, S., Curtis, S., Diez-Roux, A.V. and Macintyre, S. (2007). Understanding and representing 'place' in health research: A relational approach. *Social Science and Medicine*, 65(9), pp.1825–1838.

Cummins, S., Stafford, M., Macintyre, S., Marmot, M. and Ellaway, A. (2005). Neighbourhood environment and its association with self-rated health: Evidence from Scotland and England. *Journal of Epidemiology and Community Health*, 59, pp.207–213.

Curtis, S. (2004). *Health and Inequality: Geographical Perspectives*. Thousand Oaks, CA: Sage.

Curtis, S. (2010). *Space, Place and Mental Health*. Farnham: Ashgate.

Dadds, M.R. and Powell, M.B. (1991). The relationship of interparental conflict and global maladjustment to aggression, anxiety, and immaturity in aggressive and nonclinic children. *Journal of Abnormal Child Psychology*, 19, pp.553–567.

Di Manno, L., Macdonald, J., Youssef, G., Little, K. and Olsson, C. (2018). Psychosocial profiles of adolescents from dissolved families: Differences in depressive symptoms in emerging adulthood. *Journal of Affective Disorders*, 241, pp.325–337.

Driessnack, M. (2006). Draw and tell conversations with children about fear. *Qualitative Health Research*, 16, pp.1414–1435.

Duncan, C., Jones, K. and Moon, G. (1999). Smoking and deprivation: Are there neighbourhood effects? *Social Science and Medicine*, 48, pp.497–505.

Duncan, G.J. and Raudenbush, S.W. (1999). Assessing the effects of context in studies of child and youth development. *Educational Psychologist*, 34, pp.29–41.

Emery, C.R., Thapa, S., Do, M.H. and Chan, K.L. (2015). Do family order and neighbor intervention against intimate partner violence protect children from abuse? Findings from Kathmandu. *Child Abuse and Neglect*, 41, pp.170–181.

Emery, R.E. (1982). Interparental conflict and the children of discord and divorce. *Psychological Bulletin*, 92, pp.310–330.

Fevre, R. (2000). Identity, work and economic development. In: S. Baron, J. Field and T. Schuller (eds.), *Social Capital: Critical Perspectives*. Oxford: Oxford University Press.

Fishman, E. and Meyers, S. (2000). Marital satisfaction and child adjustment: Direct and mediated pathways. *Contemporary Family Therapy*, 22, pp.437–451.

Foley, M. and Edwards, B. (1998). Beyond Tocqueville: Civil society and social capital in comparative perspective. *American Behavioural Scientist*, 42(1), pp.5–20.

Galster, G. (2012). The mechanism(s) of neighborhood effects: Theory, evidence, and policy implications. In: Van Ham, M.Manley, D.Bailey, N.Simpson, L. and Maclennan, D. (eds.), *Neighbourhood Effects Research: New Perspectives*. Dordrecht: Springer, pp.23–56.

Gatrell, A. and Elliott, S.J. (2015). *Geographies of Health: An Introduction*. Third edition. Chichester: Wiley Blackwell.

North to carry out humanitarian missions in territories of the countries of the Global South. William Easterly, former World Bank economist, points out that 'the political crises that make the headlines today, ...have some roots in past Western treatment of peoples.... Look behind a modern-day headline and you will find the machinations of some long-forgotten colonial planner' (2008, pp. 241–242).

Child soldiers have always been present on the battlefields of adults. They fought in ancient Sparta, ancient Rome, during WWI and WWII, and took part in the American Revolution. However, in the contemporary world of global interdependencies, the stereotypical image of child soldiers appears to be a response to the demand of media viewers and readers. The recent transformation in language used to report on this subject can be observed in the example of The United Nations International Children's Emergency Fund (UNICEF). Initially, when referring to child soldiers in various reports and documents, UNICEF opted for language that would unsettle and disturb the viewers / readers, even shock them with distressing imagery of the situation of child soldiers (see: Raber n.d.). However, the agency has shifted its approach and now attempts to counter the stereotypical images of child soldiers with a language that is increasingly sensitive to children. As a result, its publications provide explanations of the situatedness, complexity and multi-faceted character of the process of child recruitment to government and opposition armed forces, as well as rebel groups. Research investigating the situation of children affected by war also attempts to give children a voice and allow them to participate in developing the reports which reveal the reality of the child who fights, flees or strives to survive. This is reflected, for example, in a recommendation included in the report 'Adult wars, child soldiers: Voices of children involved in armed conflict in the East Asia and Pacific Region': 'Ensure participation – where such participation can take place in safety and dignity – of children affected by armed conflict, including child soldiers, in all research, advocacy and programme planning activities' (UNICEF, 2001, p. 75).

The images of wartime initiation

Contemporary initiation into violence and combat has taken on many faces. It is conditioned by time, place and the socio-political specificity of a particular armed conflict or war, as well as the ideology followed by leaders and the demand for recruits. A war involving children pays off: children are flexible, they learn fast, absorb ideologies, they are more agile, smaller and inspire trust more than adults. Consequently, they require smaller temporal and material investment.

Even the first stage of the process, compulsory recruitment, i.e. drafting children as soldiers (the practice was recognised as a war crime by the International Criminal Tribunal), frequently entails acts of violence and terror. Recruited children are subjected to inhumane treatment that serves to instil new behavioural patterns oriented towards violence. They are forced to kill members of their own families, friends or co-villagers. They participate in macabre rituals with the objective of driving the young recruits towards permanent alienation and separation from families, home and communities. Moreover, having been recruited to armed forces, child soldiers experience a life of constant wartime, which entails exposure to various forms of physical and psychological violence, forced labour, sexual slavery, forced drug abuse, being tortured, being forced to torture others, and finally committing murders. However, before participating in acts of killing and torture, as Govier (2015, p. 33) stresses, children are brainwashed and threatened, and frequently also drugged or forcibly inebriated. As Karolina Janiszewska writes in the context of armed conflict in Sierra Leone, young recruits 'were prepared to service in camps. Depending on the conditions and accompanying circumstances, the training took a more or less organised form, and usually lasted between several and a dozen days, or even several months. Its primary aim was the best possible preparation of children to the new mode of life' (2013, p. 186). A crucial aspect of military and rebel training for young recruits is to break the bond that links young soldiers with their loved ones, other people from familiar backgrounds and the environment they come from, and consequently to form a new, wartime, dehumanised personality. The above is confirmed in Roberts's words: 'a soldier must learn to dehumanize other people and turn them into targets, and at the same time to cut himself off from his own feelings of caring and connectedness to the human community' (1984, p. 197). Honwana (2011) emphasises that terror is one of key tools used in the initiation of recruited children into violence. Forced to use violence towards others, their own suffering experienced through intimidation and inflicting physical pain was 'intertwined and mutually reinforcing' (Honwana 2011, p. 65). As Govier (2015, p. 33) explains, the brutality of the training leaves children terrorised and capable of participating in combat as well as serious violations, including torture, rape or plunder.

Victimised children, those who have experienced suffering and have been subjected to inhuman practices of coercion, become the most ruthless soldiers. What is more, children appear to be perfect candidates for soldiers as '...they do not fear death – neither their own nor the death of others, since unlike adults, they do not think about it...They are not afraid, as they have experienced so little, they have learnt so little, they have so little to lose. They lack experience, aspirations and dreams, obligations or duties' (Jagielski 2009, pp. 62–63). The demanding and draining military training is aimed at pushing children to the verge of physical exhaustion. This is done deliberately, as fear, fatigue and pain are favourable to the process of children's and adolescents' ideological indoctrination. Enslaving the physical body allows one to subjugate the mind. Minors are easily manipulated (cf. Honwana 2011) both through ideological measures and through coercion. In her documentation of the situation of children in Mozambique and Angola, Honwana (2011) provides a detailed description of children's initiation to the role of soldiers. Some ethnographic narratives expose the inclusion of traditional rites and rituals in modern training. One of the rituals reported by the researcher is a practice adopted by JURA (UNITA youth group).[3] The rite involves forcing young soldiers to sing and dance continuously throughout the night, which brings on psychological and physical consequences (Honwana 2011).

Child soldiers

> A child is (…) frequently a bearer of difficult truth (Orzeł 2014, p. 11).

Despite being recognisable in the media, the category of child soldiers is not as well-known as other categories (e.g. street children, children out of place, socially maladjusted children), in which, due to their difficult social or developmental situation, children – the victims, witnesses and perpetrators in contemporary conflicts and wars – are inscribed. Moreover, the portrayal of children's experiences which emerges from their narratives, scientific publications, popular science literature, court records or international organisations' reports is extremely dark and violent. Frequently, it is difficult to believe the atrocities experienced by recruited children (cf. Markowska-Manista 2012a, p. 213, Markowska-Manista 2012b, pp. 204–224.).

Research on children's and adolescents' participation in contemporary armed conflicts encompasses a number of interdisciplinary, international studies and analyses, e.g. in the fields of psychology, sociology, political studies, educational studies and childhood studies. Researchers mentioned below refer both to child soldiers' participation in wars based on the geographic criterion (case studies concentrating on particular countries and regions) as well the criteria of causes, consequences and effects (review and comparative studies). The investigations also encompass analyses oriented towards particular cases, recruitment risk factors, the complex mobilisation of children, the process of killing and attempts at their demobilisation, reintegration and social inclusion in the post-war period with the consideration of social and political changes. An analysis of academic and popular science publications written between 1994 and 2015 revealed that in the English and German literature, the prevailing subjects include those of child soldiers as a dual category of victims and perpetrators (child soldiers as victim-perpetrators and perpetrator-victims, Govier 2015; Wessells 2006), recruitment paths and children's role in wars and conflicts of longue durée (long-lasting wars and conflicts) (Cohn et al. 1994; Wessells 2006; Mazurana & McKay 2001), and legal aspects of engaging 'ordinary' (civilian) adolescents in organised practices of the torture and killing of 'ordinary people' (Drumbl 2013; Sloth-Nielsen 2013; Sloth 2018; Francis 2007; Grover 2012). Due to the emergence of new conflicts and new methods of children's and adolescents' recruitment to armed groups, including those connected with the jihadist group The Islamic State of Iraq and the Levant (ISIL), the list of issues relating to child soldiers addressed and explored in academic publications is bound to expand. It must be stressed that despite the enormous research interest in terrorism, the subject of children's participation in terrorist organisations has been largely omitted. It primarily results from the fact that these are relatively new circumstances, in new political contexts, where researchers' access to children and adults involved in terrorist activity is limited due to the threat to researchers' health and / or lives.

Another source of information on child soldiers comes from reports and publications prepared by international organisations (UNICEF, World Vision, Child Soldiers International, Human Rights Watch, Amnesty International). Former child soldiers' statements collected by NGOs,[4] included in their biographies or autobiographies, reinforce the discourse on the initiation and socialisation to violence and killing. Researchers' publications, diaries of former recruits (Beach 2010; Jal & Davies 2009), reports by international organisations, accounts of war correspondents and activists who lobby for ending child recruitment extensively document this phenomenon (UNICEF 2003; 2001). However, due to the diversity of tasks performed by children and adolescents within armed forces, it is difficult to provide an exact definition of a 'soldier' in this context, while the chaotic and dangerous circumstances inhibit the development of statistical data.

The approach which dominates in academic literature devoted to child soldiers – the victims, perpetrators and witnesses of modern wars and armed conflicts – is one of victimisation. This problem was addressed for instance by Judith Ennew, who drew attention to and opposed this dominant 'victimising' approach, not only in the case of research on childhood but also in relation to the practices of professionals working with children (more: Ennew 2009, pp. 365–378, see also McRobert's chapter on young survivors of sexual abuse, this volume). In the case of child soldiers, it is difficult to distinguish between victims and perpetrators. The status of a victim is frequently parallel to the status of a perpetrator and witness (Govier 2015). What is more, the protective status of children and the transitional status of adolescents, coupled with the awareness that children and adolescents can be relatively easily manipulated, amplify the complexity of assigning blame. A number of courts proved reluctant to assign responsibility to children for their actions in times of war (Wessells 2006, pp. 220–221).

The dyadic term 'child soldier' consists of two opposing designates. The term child, which is indistinct in its definition, is inscribed in the academic research areas of pedagogy, psychology, sociology and many other disciplines and sub-disciplines which address the issues of nurture and education connected with childhood, family or hearth and home. Its understanding is linked to the classic interpretation of care provided to an individual who has not yet reached full maturity or independence – an interpretation which is rooted in the modern-day concept of family (cf. Ariès 2010, p. 329). The term 'child' refers to the stages of a socially constructed childhood and the phases of the child's development. In other words, it refers to a period in which children require protection as they undergo a process of identity formation and socialisation. It must be stressed that concepts of the child and childhood vary across cultural contexts

(Trommsdorff 1995). They have undergone transformations both in temporal and geographical dimensions. Due to the specific cultural conditioning of countries on the African continent, the definition of a child was included in 'The African Charter on the Rights and Welfare of the Child'. The document (among other things) determined that 'a child means every human being below the age of eighteen years'.

The second element of the term, 'soldier', the protagonist of war, relates primarily to political and sociological theories of defence, the problem of military or quasi-military operations, armed activities and involvement in hostilities. Attack, oppression, violence, brutality – these are frequent associations with soldiers involved in combat. Hence the semantic conflict in the combination of the words 'child' and 'soldier'. This word-formative antinomy determines and defines a considerable number of child recruits who, either coerced or voluntarily, have found themselves among army or rebel groups in the late twentieth and early twenty-first centuries. As an act of ultimate violence which, as Carl von Clausewitz (1995) writes, sees no constraints, war opens a space for multidimensional aberrations. In truth, as Walzer (2010) points out, the logic of war lies in a constant drive towards moral extremes.

The term child soldier appears in numerous reports and resolutions of international organisations, while its definitions in various sources differ marginally (Nowakowska-Małusecka 2012, pp. 25–26). The term refers to every person under the age of 18 who has been recruited or used in any form during hostilities by an armed group. It encompasses both children incorporated into the army or militant groups, and children being part of any regular or irregular armed forces (definition based on UNICEF 1997). The term 'child soldiers' refers to children involved in combat, fighters, as well as children entangled in various interdependencies inside the recruitment system, i.e. messengers, spies, cooks, porters, girls and boys exploited for sexual services and in forced marriages, as well as any person who accompanies such groups (except family members; Child Soldiers International 2012, p. 8). A definition accepted during the 1997 conference organised by UNICEF and other global NGOs specifies that a child soldier is a boy or a girl below the age of 18 who is a member, voluntarily or by force, of any armed group, government force or fighting against some government. This includes young terrorists who were raised or even born in Europe. The events of 2015 and 2016 (terrorist attacks in Europe, conflict and civil war in sub-Saharan countries, war in Syria) show that persons under the age of 18 are not always passive victims recruited by military and quasi-military forces, including terrorist groups. At times, they take a rational decision based on the conviction that to avoid fighting is worse than combat (more: Brett & Specht 2004; Ballesteros Duarte 2010). In light of Article 1 of the Convention on the Rights of the Child, such actions imply that a person under the age of 18 is both a perpetrator and a victim of terrorism. As a consequence, such children can be subjected to strict antiterrorist legislation.

The profile and motivations of child soldiers

> The extremity of experiences blurs the specificity of childhood, dominated by the will to survive. What remains is life stripped bare (Orzeł 2014 p. 11).

There is no one single profile of a child soldier. Children's age, level of education (or lack thereof) and skills vary depending on the place they come from, the conditioning, position and role they are to play and the impact of the environment they belong to. Children's motivations to become soldiers are equally diverse (it must be stressed that the issue of motivation does not refer to children who are kidnapped, forcefully recruited or sold to fight).

The recruitment of child soldiers has many faces and takes on various forms. The narratives of former child soldiers reveal that joining an armed group often saved their lives, providing the only available source of food, shelter and protection. In certain cases, children fleeing home due to tragic family circumstances (see: Govier 2015, p. 34) joined the army or a rebel group, as doing so offered them their only chance for survival. Children without protection and care can be forced to fight an ideological war, to fight for justice and equality, for glory and honour. Having experienced a series of losses, they might also fight out of revenge. Children might perceive their active roles in armed groups as preferable to the highly restrictive roles imposed on them by the practices and values of the conservative societies they belong to[5] (see: Govier 2015, pp. 34–35). This illustrates that the reasons why a child can choose to join an armed group and decide to fight are plentiful.

Some of the recruits believe that they fight for liberation, for a dignified life for their families, communities, nations or state. Others, being raised in a culture of violence and not knowing peace or socio-political stability, seek retribution, or are driven by revenge for mistreatment, the loss of loved ones or family honour. Still others want to prove their manhood, maturity, patriotism, courage and their bond with the values of community. In other words, some wish to prove that they are capable of participating in the activities of adults (more: Bartkiewicz 2016). The effects of armed conflict and war (destruction, impoverished societies, lack of perspectives for quick social development and a dignified life) can be responsible for children's decisions to join the army or a rebel group.

> ...children see the army as the only source of support and chance for survival. Many of them are subjected to manipulation and pressure on the part of their surroundings: their families, school or religious communities. Propaganda and religious zealousness impel them to fight with a belief that they fight for the right cause. (Bartkiewicz 2016)

The climate of war and destabilisation both in the countries of Arab and sub-Saharan Africa are conducive to the practices designed to mould child soldiers into fighters, active and obedient followers of orders given by adults or peers. The widespread availability of light weapons – costing only a few dollars in some African countries, and easy for a child to handle, contribute to the prolonged and increased use of child soldiers. Even children remaining in refugee camps are not safeguarded against recruitment. As an illustration: recruitment of children under the age of 18 in Jordanian refugee camps is also encouraged by economic incentives. A UN report reveals that families receive monthly compensation from armed groups for every young fighter, or priority access to food assistance (e.g. in the Zataari camp in Jordan run by the UNHCR; Luck 2013).

It must be stressed that despite the fact that the conflicts in Sierra Leone, Mozambique, Angola, the Democratic Republic of Congo, the Central African Republic, Sudan, Darfur or Northern Uganda were broadly commented on in the media, Africa is not the only continent where children are involved in political violence. Whilst many well-known examples exist through history, including Sparta and Nazi Germany, in the 1990s, the phenomenon of child soldiers was also witnessed in Northern Ireland and the Balkans. The purposeful programming and training of children to become fighters or soldiers is part of the history of many societies and nations. Through subjecting children to ideological propaganda, adults turn them into 'cannon fodder', 'cheap labour' for wartime activities, deprive them of a safe childhood and complicate the process of maturation and entering accepted social roles.

Conclusion

In this chapter, I attempted to outline the practices of preparing children to fight as soldiers and their aides in wars and armed conflicts on the African continent, with a particular focus on the previous decade. From the point of view of social pedagogy and developmental psychology, they fit the pattern of a destructive process of educating the young generation. Due to the psychosocially and politically negative nature of this type of influence, as well as the inhumane methods and unethical tools used in this educational process, we can define this practice as a deliberate transgression of educational boundaries. It is a transgression resulting in the violation of children's rights as defined in international legislation, including the Convention on the Rights of the Child, the African Charter on the Rights and Welfare of the Child and the facultative Protocol to the Convention on the Rights of the Child on the involvement of children in armed conflict.

What is particularly difficult to understand in the phenomenon of child soldiers is their dual role. By engaging or being forced to engage in acts of killing, torture and mutilation, they appear to be both perpetrators and victims. Govier (2015) draws attention to this ambiguity, saying that a fighting child is, on the one hand, an agent performing acts of violence, and on the other hand, a person who cannot take moral responsibility for these acts since they occur in a situation the child is not responsible for. He adds that the complexity of the phenomenon can be seen in both a theoretical and practical dimension. On the theoretical level, the guilt of a (presumably innocent) child is juxtaposed with the active role of a (presumably passive) victim. In praxis, the complexity and ambiguity concern such issues as military ethics, rehabilitation and reintegration (Govier 2015, p. 35).

International law does not guarantee absolute and unchallenged protection to children involved in military activities and terrorism. The participation of underage recruits in armed conflict, in particular civil war, which has become the most common form of conflict in the twenty-first century, demands interest and consideration on the part of academics and policy makers (Lasley 2012, p. 17). Being a threat and perceived as a threat, child soldiers (perpetrators) belong (first of all) to the most vulnerable groups in times of armed conflict since they lack peer-representation to express and defend their interests (Simmons 2009).

Notes

1 The chapter is based on an article originally written in Polish and published in *Chorzowskie Studia Polityczne*.
2 UN Resolution 51/177 of 16 October 1996 recommended that the Secretary General appoint a Special Representative for Children and Armed Conflict. Following REDRESS (2006).
3 JURA, a youth organization of UNITA (The National Union for the Total Independence of Angola). The subject is addressed by Honwana (1998).
4 www.warchild.org.uk [access date: August 17, 2018].
5 Similar issues relating to children living in slightly different contexts excluded from mainstream society are addressed by Gemma Pearson in her chapter, this volume. She discusses young people living as 'street connected' youth as a preference to living in oppressive situations at home.

References

Organization of African Unity (1990) *African charter on the rights and welfare of the child*. Addis-Ababa: Organization of African Unity. Available at: https://au.int/sites/default/files/treaties/7773-treaty-0014_-_african_charter_on_the_rights_and_welfare_of_the_child_e.pdf [Accessed August 17, 2018].
Ariès, P. (2010). *Historia dzieciństwa*. Warsaw: Wydawnictwo Aletheia, p.329.
Ballesteros Duarte, M.P. (2010). *Understanding the Context of Voluntary Child Soldiers: Why did they choose to join the irregular armed forces? The Case Study of Sierra Leone*. Barcelona: Institut Barcelona d'Estudis Internacionals.
Bartkiewicz, A. (2016). *Nieletni żołnierze ofiary wojen dorosłych*. [online]. Available at: www.psz.pl/116-bezpieczenstwo/agnieszka-bartkiewicz-nieletni-zolnierze-ofiary-wojen-doroslych [Accessed August 17, 2018].
Bassompierre, L. (n.d.). *We Weren't Born Killers*. Available at: www.lucbassompierre.com/we-weren-t-born-killers [Accessed August 17, 2018].
Beach, I. (2010). Foreword. In: K. E. Dupuy, K. Peters, *War and Children: A reference handbook*. Santa Barbara: ABC CLIO. p.vii.
Brett, R. & McCallin, M. (2004). *Children: The Invisible Soldiers*. Stockholm: Save the Children Sweden.

Brett, R. & Specht, I. (2004). *Young Soldiers: Why they choose to fight.* London: Lynne Rienner Publishers.

Brocklehurst, H. (2009). Childhood in conflict: Can the real child soldier please stand up. *Ethics, Law and Society*, 4(259). pp. 259–270.

Child Soldiers International (2012). *Louder than Words: An agenda for action to end state use of child soldiers.* London: Child Soldiers International, p. 8.

Coalition to stop the use of child soldiers (2008). *Child Soldiers, Global Report 2008.* Available at: www.hrw.org/legacy/pub/2008/children/Child_Soldiers_Global_Report_Summary.pdf [Accessed August 17, 2018].

Cohn, I., & Goodwin-Gill, G.S. (1994). *Child soldiers: The role of children in armed conflict.* New York: Oxford University Press.

Drumbl, M.A. (2013). Transcending victimhood: Child soldiers and restorative justice. In: T. Bonacker & C.J.M. Safferling (eds.), *Victims of International Crimes: An Interdisciplinary Discourse.* The Hague: TMC Asser Press, pp. 129–130.

Easterly, W. (2008). *Brzemię białego człowieka.* Warsaw: PWN, pp. 241–242.

Ennew, J. (2009). *The Right to Be Properly Researched: How to Do Rights-Based, Scientific Research with Children.* Bangkok, Thailand: Black on White Publications.

Francis, D.J. (2007). 'Paper protection'mechanisms: Child soldiers and the international protection of children in Africa's conflict zones. *The Journal of Modern African Studies*, 45(2), pp. 207–231.

Grover, S.C. (2012). *Child soldier victims of genocidal forcible transfer: Exonerating child soldiers charged with grave conflict-related international crimes.* Berlin: Springer Science & Business Media.

Govier, T. (2015). *Victims and Victimhood.* Toronto: Broadview Press. p. 34.

Honwana, A. (1998). *Okusiakala ondalo yokalye – 'Let us light a new fire': Local knowledge in the post-war healing and reintegration of war-affected children in Angola.* Angola: Christian Children's Fund. Reproduced and available at: www.child-soldiers. org/psycho-social/english [Accessed: January 29 2019].

Honwana, A. (2011). *Child Soldiers in Africa.* Philadelphia: University of Pennsylvania Press.

Jal, E., & Davies, M.L. (2009). *War child: A child soldier's story.* New York: Macmillan.

Jagielski, W. (2009). *Nocni wędrowcy.* Warsaw: Wydawnictwo W.A.B. pp. 62–63.

Janiszewska, K. (2013). Sierra Leone–wojna z dziećmi w roli głównej: Refleksje. *Pismo naukowe studentów i doktorantów WNPiD UAM*, 8, p.186.

Kaplan, R.D. (1994). The coming anarchy: How scarcity, crime, overpopulation, tribalism, and disease are rapidly destroying the social fabric of our planet. *The Atlantic*, February. www.theatlantic.com/magazine/archive/1994/02/the-coming-anarchy/304670/ [Accessed January 27, 2019].

Lasley, T.C. (2012). *Creed vs. Deed: Secession, Legitimacy, and the Use of Child Soldiers.* Theses and Dissertations: Political Science. Available at: https://uknowledge.uky.edu/polysci_etds/2/ [Accessed August 17, 2018].

Lee, A.J. (2009). *Understanding and Addressing the Phenomenon of 'Child Soldiers' (Working Paper Series No. 52).* Oxford: Department of International Development, University of Oxford.

Liebel, M. (2015). *Kinderinteressen: Zwischen Paternalismus Und Partizipation.* Weinheim: Beltz Juventa, p. 300.

Luck, T. (2013). As Syrian rebels' losses mount, teenagers begin filling ranks. *Washington Post*, August 24. www.washingtonpost.com/world/middle_east/as-syrian-rebels-losses-m

ount-teenagers-begin-filling-ranks/2013/08/24/2bdbdfea-0a8f-11e3-9941-6711ed662e71_ story.html?noredirect=on&utm_term=.cecd205bbd51 [Accessed January 27, 2019].

Machel, G. (2002). *The Impact of Armed Conflict on Children: United Nations, August 1996, A/51/306.* Available at: www.unicef.org/graca/a51-306_en.pdf [Accessed August 17, 2018].

Markowska-Manista, U. (2012a). Ofiary, sprawcy, świadkowie. Socjalizacja dziecka doświadczonego piętnem konfliktu, wojny i ludobójstwa w zróżnicowanych etnicznie regionach Afryki Subsaharyjskiej. In: H. Klarkowska–Grzegołowska (ed.), *Agresja, socjalizacja, edukacja. Refleksje i inspiracje.* Warsaw: Wydawnictwo APS, pp. 159–180.

Markowska-Manista, U. (2012b). Dzieci – ofiary, świadkowie i sprawcy ludobójstw w krajach Afryki Subsaharyjskiej. In: A. Bartuś, & P. Trojański (eds.), *Auschwitz a zbrodnie ludobójstwa XX wieku, Państwowe Muzeum Auschwitz-Birkenau.* Oświęcim: Fundacja na rzecz MDSM, pp. 204–224.

Mazurana, D., & McKay, S. (2001). Child soldiers: What about the girls? *Bulletin of the Atomic Scientists*, 57(5), pp. 30–35.

Mcloughlin, C. (2009). *Topic guide on fragile states.* Birmingham: International Development Department, University of Birmingham. Available at: www.gsdrc.org/ docs/open/con67.pdf [access date: August 17, 2018].

Mendelsohn, M., & Straker, G. (1998). Child soldiers: Psychosocial implications of the Graça Machel/UN study. *Peace and Conflict: Journal of Peace Psychology*, 4(4), pp. 399–413.

Nowakowska-Małusecka, J. (2012). *Sytuacja dziecka w konflikcie zbrojnym: Studium prawno-międzynarodowe.* Katowice: Wyd. Uniwersytetu Śląskiego, pp. 25–26.

Orzeł, O. (ed.) (2014). Introduction. In: *Dziecko żydowskie w czasach Zagłady. Wczesne świadectwa 1944–1948. Relacje dziecięce ze zbiorów Centralnej Żydowskiej Komisji Historycznej.* Warsaw: ŻIH, p. 11.

Poddler, S. (2011). Neither child nor soldier: Contested terrains in identity, victimcy and survival. In: A. Özerdem & S. Podder (eds.), *Child Soldiers: From Recruitment to Reintegration.* Hampshire/New York: Palgrave Macmillan, p. 146.

Prucnal, M. (2012). *Ochrona dzieci przed uczestnictwem w działaniach zbrojnych we współczesnym prawie międzynarodowym.* Warszawa: Biuro Rzecznika Praw Dziecka.

Raber, N. (n.d). *UNICEF Perspectives on African Child Soldiers.* Available at: https:// nicholasraber.wordpress.com/school-2/major-projects/paper-unicef-perspectives-on-a frican-child-soldiers/ [Accessed August 17, 2018].

REDRESS (2006). *Victims, perpetrators or heroes? Child soldiers before the International Criminal Court, September 2006.* London: The REDRESS Trust. Available at: www.redress.org/downloads/publications/childsoldiers.pdf [Accessed date: August 17, 2018].

Roberts, B. (1984). The death of machothink: Feminist research and the transformation of peace studies. *Women's Studies International Forum*, 7(4), p.197.

Simmons, B.A. (2009). *Mobilizing for Human Rights.* Cambridge: Cambridge University Press.

Sloth, J.J. (2018). Monitoring and implementation of children's rights. In: Kilkelly, U., Liefaard, T. (eds.) *The International Human Rights of Children.* Singapore: Springer.

Sloth-Nielsen, J.J. (2013). Deprivation of liberty 'as a last resort' and for the 'shortest period of time': How far have we come? And can we do better? *Journal of Criminal Justice*, 26(3), pp. 316–337.

Trommsdorff, G. (ed.) (1995). *Kindheit und Jugend in verschiedenen Kulturen. Entwicklung und Sozialisation in kulturvergleichender Sicht.* Weinheim/München: Juventa Verlag.

UNICEF (1997). *Principles and Best Practices on the Recruitment of Children into the Armed Forces and on Demobilization and Social Reintegration of Child Soldiers in Africa.* Cape Town: UNICEF. Available at: www.unicef.org/emergencies/files/Cape_Town_Principles(1).pdf [Accessed August 17, 2018].

UNICEF (2001). *Adult wars, child soldiers: Voices of children involved in armed conflict in the East Asia and Pacific Region.* Bangkok: UNICEF. Available at: www.unicef.org/evaldatabase/files/EAPRO_2001_AdultWars.pdf [Accessed August 17, 2018].

UNICEF (2002). Fragment of a statement by a 17-year-old girl, Elisa, during a special session of the UN General Assembly. Available at: www.unicef.org/specialsession/voices/ [Accessed date: August 17, 2018].

UNICEF (2003). *Guide to the optional protocol on the involvement of children in armed conflict.* New York: Coalition to Stop the Use of Child Soldiers, UNICEF. Available at: www.unicef.org/sowc06/pdfs/option_protocol_conflict.pdf

UNICEF (2016). *No place for children: The impact of five years of war on Syria's children and their childhoods.* Available at: www.unicef.de/blob/106970/ef5ba0af8e768eacb0baefe0ca13d3ae/unicef-bericht-no-place-for-children-2016-03-14-data.pdf [Accessed August 17, 2018].

United Nations (2000). *Optional protocol to the convention on the rights of the child on the involvement of children in armed conflict, General Assembly Resolution 54/263*, adopted May. Available at: www.ohchr.org/en/professionalinterest/pages/opaccrc.aspx [Accessed January 28, 2019].

von Clausewitz, C. (1995). *O wojnie*, trans. A. Cichowicz, L. Koc. Lublin: Test.

Walzer, M. (2010). *Wojny sprawiedliwe i niesprawiedliwe.* Warsaw: PWN, p. 66.

Wessells, M.G. (2006). *Child Soldiers: From Violence to Protection.* Harvard: Harvard University Press, pp. 220–221.

Zwoliński, A. (2009). *Biedy Afryki.* Kraków: Petrus, p. 217.

Part II
Work, education and activism

6 Other(ed) childhoods

Supplementary schools and the politics of learning

Helen F. Wilson, Erica Burman, Susie Miles and Saskia Warren

Chapter summary

Supplementary schools are an important part of the UK's educational heritage yet their contributions are frequently overlooked. Run by communities, often by volunteers, supplementary schools offer out of hours education for black and ethnic minority children and young people to support language tuition, cultural and spiritual learning, religious study, and, in some cases, additional support in core academic subjects. In outlining the socio-political histories that have shaped the landscape of supplementary provision, this chapter demonstrates how supplementary schools have long responded to racism and the effects of structural violence whilst creating an important space for cultural preservation. The chapter connects supplementary schools to critical debates on the geographies of education, and interrogates the distinctions that are drawn between the mainstream and its supplementary 'other'. In doing so, we outline an important lens through which to examine the production of educational disadvantage and the practices of resistance that have provided alternative learning spaces in multicultural contexts that continue to be shaped by the legacies of colonialism. The chapter further underlines the importance of attending to the differences that exist between supplementary schools that face very different challenges and opportunities.

Introduction: going for gold...

In 2015, on a cold December evening, around 40 people gathered together for the annual meeting of supplementary schools in Manchester, UK. Two speakers from two schools had been invited by the local council to give a talk on their experiences of 'going for gold', the highest award granted by the National Resource Centre for Supplementary Education (NRCSE). Each school had brought their portfolio along as a demonstration of the weight of work that had gone into their application. Lesson plans, assessment examples, and pieces of school work, detailed accounts of staff development and training, pictures of classrooms and facilities, checklists and references, all arranged meticulously in plastic wallets, were held up and passed around. The speakers detailed the investments that had been made in their school facilities and equipment, the language classes for adult learners that were bringing in additional revenue, and the contributions from parents that had allowed them to invest in new media and software.

When the discussion was opened up to the floor there were multiple questions from other school representatives: how can you bring in money if the language you teach is not considered to be economically profitable? What if you don't have any income from alternative sources or are working out of a temporary venue? What if you are working with children with complex issues or children that have yet to be granted the legal right to remain? What happens if you have limited staff with no access to training? What if parents can't afford to pay a fee or if you can't afford enough for exercise books, let alone ICT equipment? Whilst all the people gathered in the room were brought together on the basis of a shared investment in supplementary schools, the differences within the room were palpable. For some, the awards system was out of reach and at the bottom of a long list of far more urgent priorities.

It has been repeatedly noted that supplementary schools face multiple obstacles (RSA, 2015). The opening account depicts a meeting in which schools were brought together by a local council for support: to share advice, discuss dilemmas and raise issues. By showcasing two schools that had 'gone for gold' and achieved it, the meeting is evidence of the ongoing processes through which supplementary schools are encouraged to assume, or work towards, various forms of formalisation through accreditation. In this context – and according to the council – supplementary schools are defined as those that offer out of hours education for black and minority ethnic children. These schools are run voluntarily by the community and provide support in language tuition, cultural and spiritual learning, religious study, and, in some cases, additional support in core academic subjects such as English, maths and science. This support is offered in addition to mainstream education, and normally covers forms of tuition that cannot be accommodated in mainstream classrooms. The charity that administers the awards system in question, The National Resource Centre for Supplementary Education (NRCSE), is the only organisation that is specifically dedicated to supporting and advising supplementary schools in England. Based in London, the organisation runs an awards scheme that grants bronze, silver, and gold quality framework awards as a means of encouraging schools to raise standards in teaching and learning, improve safeguarding, staff development and good financial management, as well as develop partnerships with other institutions. The bronze award is designed to identify schools that have demonstrated that they have basic management procedures in place (such as staff training or background checks), in order to keep children safe in accordance with local authority and Ofsted statutory requirements for safeguarding. The silver and gold awards, however, require more detailed evidence of staff training, teaching innovation, partnerships, and 'excellent' learning environments. The successful completion of such awards is presented as an opportunity to gain further credibility and open up avenues for more funding as a result, which is crucial at a time when schools are struggling to secure revenue.

As is evident in the opening account, this system of formalisation and accreditation can take considerable time and resources and has the capacity to exacerbate some of the inequalities and differences that exist across a sector that already faces considerable challenges. Using this as our starting point and focusing predominantly on the UK, our chapter focuses on this often neglected and regularly misunderstood sector of education at a time when resources and support for such schools are stretched increasingly thin. In so doing, we draw attention to a form of education that has not only been maligned politically, but overlooked in academic research. As we contend, this is a significant oversight, as supplementary schools are an important lens through which to understand structural and educational inequalities, practices of resistance and struggle – and their associated politics – and the complex geographies of migration and belonging. In drawing attention to a little understood form of education, we further underline the disparities that exist between schools which experience different challenges and opportunities, making any attempt to treat supplementary schools as a homogenous and stable category of education problematic.

The section 'What are supplementary schools' offers an introduction to supplementary schooling and its socio-political histories to demonstrate how, far from simply responding to needs that cannot be accommodated within mainstream classrooms, supplementary schools have long been associated with struggles against educational inequality. Having noted how supplementary schools respond to forms of structural discrimination, the section 'The supplementary "other"' examines how supplementary schools are ambiguously positioned in relation to mainstream education and alternative forms education, tracing the ways in which supplementary schools have been routinely 'othered'. As part of this work, we turn to critical geographies of education to outline why supplementary schools are so important to current debates (cf. Basu, 2013; McCreary et al., 2013; Pini et al., 2017). The next section, 'The differences within...', builds on these points to outline the contemporary challenges faced by supplementary schools and to further address the disparities that exist between them and yet are routinely overlooked. The chapter finishes by underlining the challenges that exist when working within and against educational categories, whilst noting the importance of supplementary schools for a wide variety of concerns beyond education.

What are supplementary schools?

Supplementary schools offer 'out of hours' education, which, in its broadest definition, can encompass breakfast clubs, homework classes, Saturday schools, and forms of religious schooling (such as Christian Sunday schools or Islamic Madrassas) that are variously considered to be attended *in addition* to mainstream schools (Clennon, 2014). Importantly, a distinction should be made between 'enrichment' activities, such as those promoted by the UK government (c.f. Holloway and Pimlott-Wilson, 2014), and supplementary forms of schooling that are community-led and often run by volunteers. Generally speaking, supplementary schools are those schools that provide additional learning support for black and ethnic minority children. Whilst there is considerable variability between schools, they variously offer language training, religious tuition, and multifarious forms of cultural and spiritual learning, as well as support in core academic subjects and sport. As a consequence, they have long been associated with struggles against educational inequality, racism, and the pathologisation of migrant communities.

It is estimated that there are 3,000–5,000 supplementary schools in England alone, and that up to 18% of children and young people from black, Asian, and ethnic minority backgrounds attend supplementary schools at some point in their childhood (Evans and Gillan-Thomas, 2015). On the NCRSE database alone, the range of schools is noteworthy. It includes schools specifically catered for refugee and asylum-seeker children, which run English language classes and offer support in academic subjects, as well as crafts, sports and baking; Persian schools for second generation Iranian children focused on language learning, maths and science; black supplementary schools that give priority to black history studies alongside support in core academic subjects; German language schools that focus on promoting awareness of the culture and values of German-speaking countries; and Russian schools that celebrate dual heritage and run language classes for children who have a Russian-speaking ethnic background. These schools take on a variety of forms including: weekend and evening classes; classes that map onto the UK's key stage curriculum and/or support study for foreign language qualifications; schools that follow other national curriculums so that children can gain overseas qualifications and strengthen diasporic connections; homework clubs that support children in their mainstream studies, and so on. A considerable number of schools place emphasis on supporting dual heritage and multi-culture and also offer support for parents who struggle to engage with the mainstream school system.

The emergence of supplementary schools is thus closely connected to histories of migration, and there is a particularly long history of supplementary schooling across North America and the UK. At the same time, new forms of migration in 'new' immigration countries has seen a spike in interest in the forms of learning and support that they can offer different communities (c.f. Aurini, 2013; Cook, 2013; Tereshchenko and Grau Cárdenas, 2013; Zhou, 2008). Socio-political histories of supplementary education in the UK chart the emergence of supplementary schools as a response to racism and educational discrimination on the one hand, and concerns pertaining to the loss of identity on the other, either in the face of deliberate attempts to assimilate or marginalise children, or as a result of a limited ability to accommodate diverse needs in the classroom (Wilson and Warren, 2019, in press). The appearance of supplementary schools is frequently connected to migrations that occurred in the aftermath of the Second World War, particularly from the Commonwealth. However, some of the earliest examples of supplementary schools have been traced back to the early 1800s, when schools were established by Irish, Russian, and Italian migrants, with Chinese schools established in the second half of the century in order to support the children of Chinese dock workers (Conteh, 2007). Burman and Miles (2018) also note that such schools weren't always necessarily supplementary in nature. For instance, the establishment of Jewish schools in Manchester was likely to have preceded the establishment of formal education for all children in the UK in 1871. The provision of education by the Jewish community was not supplementary to mainstream education, but rather played a significant role in agitating for the introduction of mainstream schooling for Jewish children, alongside which Jewish schools later came to exist as 'supplementary' (ibid.).

Across the literature, then, there are no agreed or coherent histories of supplementary schooling, with different commentators identifying different start points and rationales for its establishment as something akin to a sector. Importantly, it is not our intention to offer coherence to these histories but, rather more simply, to note the deep-rooted connections between supplementary schools and migration. That said, it is also worth underlining the existence of other forms of identity-based supplementary schooling. For example, Gerrard's (2011) study of early twentieth century British socialist Sunday schools, which predominantly catered for working class children, is testament to forms of supplementary schooling that were organised around political forms of identification, but that were also used to challenge dominant conceptions of 'culture'. In attending to questions of migration, it is also important to note that migrations take different forms and that the conditions that gave rise to early forms of 'supplementary education' were different from later forms. Schools that have been established as a result of forced migration from war-torn countries, for instance, face very different challenges to those established by economic migrants as a result of rather more privileged circumstances.

Whilst the landscape of supplementary schooling is historically fragmented, Myers and Grosvenor (2011) note the importance of connecting the experiences of different 'colonial minorities' as a means of documenting how supplementary schools emerged as part of a wider history of anti-racist struggle and self-determination (see also Andrews, 2013; Reay and Mirza, 1997; Sneddon and Martin, 2012; Wei, 2006). For instance, they note how Irish supplementary schools in England were established in the second half of the twentieth century in response to a variety of factors, including: concerns over a loss of meaningful attachment to Irishness in second generation migrants; the production of the Irish community as 'suspect' in relation to the bombing campaign of the IRA; the lack of space afforded to the articulation of Irish identities in English Catholic schools; and a mainstream curriculum that persistently disappeared the violences of empire. In response, Irish schools offered Irish language lessons, taught Irish history and dance, and provided the space for exhibitions and ceilidhs. At the same time, the black supplementary school movement, established in the 1960s, responded to racism and the 'codes and values of white hegemony' in mainstream education, cultures of low expectation that frequently categorised black children as 'slow learners', and the absence of black history in the curriculum (Myers and Grosvenor, 2011: 506–509; see also Andrews, 2013). Black supplementary schools were thus intended to provide a space free from racism where children could receive educational support in mainstream subjects, as well as lessons in black history and culture as an antidote to colonial forms of education.

Some schools were – and still are – more politically active than others, and there is a long history of contrasting ideologies between schools that had a more critical stance towards mainstream schooling and its curriculum, and those that sought to better support it (Andrews, 2013). However, regardless of their differences, these histories demonstrate how supplementary schools do more than simply offer tuition in those things that cannot be accommodated in the classroom, such as diverse forms of language learning or dance. In many cases schools explicitly responded to the erasure of migrant histories and the institutional violences and racism of mainstream schooling to create safe and comfortable spaces in which children could learn and receive support by people attuned to their needs. It is to these concerns that we now turn.

The supplementary 'other'

As Wilson and Warren (2019, in press) have argued, supplementary schooling should be of paramount concern to critical geographies of education. In honing in on the various ways in which marginality and disparity are (re)produced through education, this body of work has concerned: the governmentality of educational rights and the mechanisms through which underachievement is constructed (Watkins and Noble, 2013); the marketization of education, legacies of disinvestment and its disproportionate effects on ethnic minorities (Lipman, 2013); attacks on teachers unions; the exclusionary and discriminatory practices that occur within educational spaces (McKinnon et al., 2017; Robbins, 2008; Riley and Ettlinger, 2011; Wilson, 2013); and the development and mitigation of 'otherness' (von Benzon, this volume). As many have argued, the stratification of educational opportunities has an unequal impact on already marginalised communities and thus demands a political economy of education that foregrounds race and the experiences of working-class parents and students. These experiences are not only repeatedly rendered invisible in education policies, but rooted in specific experiences of (post)colonialism, imperialism, and the diminishment of state welfarism (Lipman, 2013: 5; Katz, 2017).

The history of supplementary schooling has demonstrated a deep concern with the unequal impact of educational opportunities and racist cultures of low expectation that have not only hindered the educational chances of some children, but have created decidedly hostile spaces within which to work. As Noble (2017) has argued, long histories of fashioning some children as 'educational problems' – children that are variously considered to be underachievers, truants, or prone to poor behaviour – tells us less about children and more about institutional inequalities, 'schemas of perception' (Watkins and Noble, 2013) and processes of categorisation. There is a significant body of work dedicated to exposing and examining the consequences of such cultures of expectation through a variety of cases, to which the experience of supplementary school learners should be added. This includes research on anti-black racism, the pathologisation of people of colour (Lipman, 2011; Robbins, 2008), and detailed studies that explicate how exploitative labour conditions are maintained by dispossessing immigrant students of 'potential futures' through racialised exclusions from education (Cahill et al., 2016: 121). In examining the production of 'less-than-average' students across a variety of contexts (Farrales, 2017), this research continues to reveal how immigration procedures, education policies and institutional practices work together to shape educational trajectories, and defer, delay, and deny, the academic achievement of minority students. This is a concern that is also present in research focused on the generation of 'moral panics' and circulating discourses of deviance. This is especially the case with the current emphasis on monitoring school children for signs of 'extremism' (Arthur, 2015; Shain, 2011), where it has been shown that the disadvantages suffered by some Muslim children are consistently overlooked and rendered invisible by concerns that they may pose a 'threat' to the British state (Holmwood and O'Toole, 2017).

Whilst there is a burgeoning literature on forms of educational discrimination, supplementary schools remain absent from much of these discussions despite being central to a variety of forms of 'resistance' that have sought to bring such hidden systems of inequality into view. As Clennon (2014: 34) has argued, the notion of resistance or activism in the context of education should invite reflection on what it is, exactly, that is being resisted. Whilst acknowledging that there is consensus around the fact that structural inequalities take an 'infinite number of forms', it is not always clear what the aims of resistance are. For instance, is it about liberation from structural inequalities, which are often felt acutely by different communities on the ground, or is it about questioning the purpose of education more broadly? Whilst the first question might be about resisting the racist production of educational problems, the second question might concern the extent to which state-led education may prohibit individual empowerment or may come into conflict with other expectations of education (Clennon, 2014: 34–35).

Given the historic and contemporary presence of various forms of resistance in supplementary schools (Dove, 1993), there are important connections to be made between supplementary schools and alternative forms of education that are established on the basis of similar concerns. Kraftl's (2013) work on forest schools, Steiner schools, and 'democratic schooling', to name just a few, documents a variety of educational contexts where some of the normative structures of mainstream schooling have been challenged or replaced. The normative structures of concern variously cover age-based segregation, a lack of opportunity to learn from the everyday or the environment, too much emphasis on league tables and narrow forms of assessment (Finn, 2016), and the reproduction of social norms. Whilst these examples centre most fully on questions concerning the status and form of education that is offered by the mainstream, there are other examples that deal more explicitly with resisting structural inequalities, especially in relation to race and culture (Clennon, 2014). The establishment of alternative community schools in North America, for example, reveal long histories of anti-racist struggle against colonial oppression and cultural assimilation. Schools, such as those established by indigenous communities, not only responded to racist cultures of low expectation, trauma and abuse, which consistently disempowered minority children (Hamilton and Ture, 1967; hooks, 2003; McCarty, 2002), but in many cases also resisted Euro-American colonial curriculums as a means to achieve greater self-determination (Davis, 2013).

Whilst alternative schools have very different histories, and are shaped by different power structures and forms of support, they all respond to dominant forms of education that are taken to be the norm. As Clennon (2014: 36) argues, because 'the dominant view is presented as being neutral and the de facto starting point for debates about education' we see a 'deleterious effect on the voicing of different views' (ibid.). Indeed, whilst Kraftl (2013: 4) has demonstrated the rather more complex connections that exist between alternative education and a variety of 'mainstreams' so as to undermine the binaries that are routinely created between 'mainstream education' and other forms of learning, it is worth interrogating the status of 'supplementary' forms of education more fully (see also Dixon-Román, 2010). As Burman and Miles (2018) have argued, a critical interrogation of the language used to describe and name 'supplementary' forms of education can reveal a lot about the cultural politics of mainstream schooling. Drawing on Derrida's (1976) notion of 'the supplement', they argue that the 'designation of "supplementary" works precisely to confirm the status of that which it is positioned to supplement' (Burman and Miles, 2018: 11). Such a designation ensures that the exteriority or 'subaltern instance' (Derrida, 1976: 145) of supplementary education is reified in relation to the so-called 'mainstream', making it somehow secondary to that which is considered to be the original. This offers a framework for interrogating the distinctions that are drawn between the mainstream and its supplementary 'other'. The 'supplementary' retains an element of ambiguity as to whether it functions as an addition, or whether it 'adds only to replace' (Derrida, 1976: 144), allowing it to be drawn into the original 'by the mark of an emptiness' (ibid.). Importantly, for Burman and Miles (2018), these distinctions between the mainstream and its other reflect and sustain the marginal status of supplementary schools and further produce material inequalities. Yet, this framing of the 'marginal' can also recognise attempts to resist incorporation into normative structures; the marginal status of the supplementary also becomes a site of refusal (hooks, 1989).

Whether the site of refusal or the production of material inequality, the designation of supplementary can work to 'suppress' the diversity of supplementary schools by foregrounding what they hold in common: their relationship to the mainstream. The penultimate section of this chapter thus engages with some of this diversity to bring it into view and underline its critical importance.

The differences within...

Because supplementary schools are so regularly defined by their relation to the mainstream – or their 'supplementary' status – the enormous diversity of supplementary schools is often rendered invisible and collapsed into a homogeneous category. Supplementary schools vary considerably in size, status, and resources, and as we have already indicated, they can have vastly different aims, intentions and relations to mainstream education (Andrews, 2013). Whilst some focus almost entirely on the maintenance of heritage in a context where needs cannot be met in the classroom, others respond more explicitly to forms of discrimination, or offer alternatives to mainstream education, even whilst they are attended in addition. Here, we want to focus specifically on the inequalities that exist between supplementary schools in order to raise pertinent questions concerning the disparities that are overlooked.

The meeting with which we began the chapter was organised by Manchester City Council as part of a wider programme of support for supplementary schools. What is notable in this account was the gap between those schools that had 'gone for gold' or were perhaps working towards it and those that could not foresee a circumstance in which they would ever be able to achieve it. Given the presentations by the two outstanding schools in question, the message was fairly simple: generate revenue; be organised; and invest in facilities. There was no indication of the diversity in the room.

Some of the schools in the room were more likely to be teaching children from economically deprived areas, whose parents couldn't afford to make contributions. Whilst some schools had formal partnerships with mainstream schools or other institutions that had provided them with a venue, others had failed to secure such 'lucrative' partnerships, were in temporary accommodation, or paying rent that was unsustainable (RSA, 2015). Funding challenges have been a common focus of our discussions with school representatives. Some of the most discernible differences were those that existed between long-settled migrant communities and relatively new arrivals, particularly those that had arrived as a result of forced migration, such as those from Syria or Somalia. These recent arrivals were trying to provide some form of continuity at a time of otherwise significant upheaval, and were still negotiating a variety of bureaucratic systems, were working with children with complex needs, as well as dealing with a variety of hostile systems that left them with little social capital or stability.

For many schools, a lack of funds leaves them entirely reliant on voluntary staff who give up their evenings and weekends, often in addition to full-time jobs elsewhere, making schools vulnerable to shortage and inconsistent delivery. A recent RSA (2015) report noted 'volunteer fatigue' as a key concern for many supplementary schools. A frequent turnover of staff makes for an unstable learning environment and a constant sense of precarity. Whilst some schools had little or no financial support to work with, others, including the schools that had 'gone for gold', receive support from other national governments as a means to support learning that can strengthen diasporic connections, maintain cultural and linguistic continuity, and potentially facilitate return migration in the future (Creese, 2009; Lytra and Martin, 2010). The delivery of language classes for the wider community was noted as one potential avenue for income, but it has long been recognised that whilst some languages are considered economically profitable due to their worldwide status – and are thus in demand (Francis et al., 2009) – other, so-called 'community languages' have often been regarded as a minority interest and are thus unlikely to generate enough interest from the wider community to be economically profitable (Kenner and Ruby, 2012). Further still, some communities continue to fall under the scrutiny of a variety of different national policies that have treated bilingual competence as 'profoundly suspect' (Lytra and Martin, 2010).

In this vein, surveillance and political scrutiny is another important concern (RSA, 2015). In 2015, supplementary schools were placed under public scrutiny when the government announced new plans to tackle what it described as 'illegal schools'. By describing schools that were teaching religious intolerance, and focusing largely on Islamic Madrassas and Jewish Yeshivas, supplementary schools were conflated with so-called 'illegal schools' that were functioning as a replacement for mainstream education (Sellgren, 2015). The prime minister's announcement not only placed supplementary schools in the headlines, but also connected them to debates on 'extremism' and segregation, and raised concerns about spaces of education where divisions were deliberately fostered. Such public scrutiny reduced the sector to a limited set of divisive discussions around migration, nationhood, surveillance and integration, whilst creating confusion as to their constitution. These plans to tackle 'illegal schools' went hand-in-hand with policies for producing conceptions of belonging that are somehow 'more aligned' with national values as part of wider scrutiny pertaining to loyalty to the British state (Shain, 2011; Sellgren, 2015; see Wilson and Warren, 2019, in press for a fuller discussion). In the aftermath of these announcements, schools reported the collapse of their rental agreements, demonstrating the damaging and uneven consequences of political scrutiny, changing policy narratives and contemporary racisms (NRCSE, 2015). As the history of supplementary schooling has demonstrated, surveillance of 'suspect communities' has been a continual feature, even if the communities under surveillance have changed (consider, for example, the political scrutiny of Irish schools in the 1960s).

In summary, the differences in capacity, resources and political support outlined above underline the importance of not only including supplementary schools within critical geographies of education concerned with the uneven geographies of provision, but addressing the inequalities that exist between them.

Conclusion: other(ed) childhoods

As we have suggested, research on supplementary schools is remarkably scarce, leading the Paul Hamlyn Foundation (2015) to suggest that critical and engaged research on supplementary schools is urgent. Not only are supplementary schools marginalised in academic research, but previous studies have found that supplementary school learners are themselves frequently marginalised, suffering higher than average levels of educational disadvantage (Strand, 2007). The urgency noted by the Paul Hamlyn Foundation especially relates to the difficulties that have been faced by supplementary schools that have sought to document their value and contributions in a context where it has been felt that the government has routinely failed to recognise their contributions (RSA, 2015). This is a challenge that has been noted by alternative forms of education more broadly (Kraftl, 2013). This lack of acknowledgement further coincides with growing frustrations concerning diminishing funds and the withdrawal of government support. Whilst local authority support has been historically patchy and uneven across the country, cuts to budgets, reallocation of staff and increased political scrutiny has, for some schools, further exacerbated their precarity and limited their options.

The diversity of the supplementary landscape can present further challenges for research. As we have noted, 'supplementary education' can refer to a wide range of activities, some of which might be variously considered to be forms of 'alternative education'. This lack of coherency and agreement around what supplementary schools look like should be further placed within the context of a bewildering diversification of educational forms, schools, and markets, which makes it increasingly difficult to research across cases and international boundaries as school categories change and slide. In this chapter we have prioritised supplementary schools that run in addition to mainstream schooling and that provide out of hours learning for black and minority ethnic children and young people, so as to focus on the intersections between migration, educational discrimination and cultural provision. By connecting supplementary schools to critical debates on the political economy and geographies of education, we suggest that supplementary schools offer an important lens through which to consider the structural inequalities that shape educational disadvantage, alongside the practices of resistance and struggle that have provided alternative spaces for learning in multicultural contexts that continue to be shaped by the legacies of colonialism (Wilson and Warren, 2019, in press).

Relatedly, it is worth noting that as ethnic minorities continue to be underrepresented in higher education, particularly in Russell Group universities (Bhopal, 2018), it has been suggested that supplementary schools might have a role to play in supporting black and minority ethnic pupils into higher education. Recognising the important role that supplementary schools play in improving educational outcomes is vital. However, this presents two problems that are worth noting. First, accepting this role does not begin to address the exclusionary practices that disadvantage applicants from black and minority ethnic backgrounds in the first place. This is a cause of concern for Andrews who has argued that too many supplementary schools are paradoxically orientated to the deficits of mainstream education by focusing on how to help children achieve success despite discrimination, rather than 'indicting the racism in the classroom' (Andrews, 2014: 68). Second, given the pressure that many supplementary schools are under, extending their remit risks volunteer burnout, a sentiment that was palpable in the council meeting with which we opened this chapter.

Finally, whilst this chapter has outlined the importance of addressing the lack of attention granted to supplementary schools in academic research and has specifically situated the discussion within critical geographies of education, it is pertinent to consider how a focus on supplementary schools brings a variety of other concerns – beyond education – into view. For instance, in noting the forms of surveillance that some communities have been subjected to, particularly in relation to the UK government's anti-extremism agenda, we note how supplementary schools speak back to concerns about forms of 'everyday bordering' and the contested politics of assimilation and nationhood (Yuval-Davis et al., 2018). Indeed, supplementary schools are of importance to critical debates on the complex geographies of migration more broadly, whether the geographies of lived multiculture (Vertovec and Wessendorf, 2010; Wilson, 2015), urban geographies of migration (Darling, 2017), the religious pluralities of cities or the diverse religiosities of young people (Dwyer et al., 2013; Hopkins et al., 2011). In short, it is not only pertinent that the contributions of supplementary schools are better addressed, alongside the struggles and forms of discrimination that are experienced by supplementary learners (especially at a time when schools are being increasingly incorporated into a variety of formal structures), but that academic debates learn from supplementary schools and their diverse and fraught experiences of responding to educational inequality and the complex geographies of migration.

References

Andrews, K. (2013). *Resisting Racism: Race, inequality, and the black supplementary school movement*, London, Institute of Education Press.

Andrews, K. (2014). Resisting racism: The black supplementary school movement. In: Clennon, O. (eds) *Alternative Education and Community Engagement: Making education apriority*, Basingstoke, Palgrave Macmillan.

Arthur, J. (2015). Extremism and neo-liberal education policy: A contextual critique of the trojan horse affair in Birmingham schools, *British Journal of Educational Studies*, 63, 311–328.

Aurini, J. (ed) (2013). *Out of the Shadows: The global intensification of supplementary education*, Bingley, Emerald Group Publishing.

Basu, R. (2013). Multiplying spaces of subalterity in education: From ideological realms to strategizing outcomes, *Canadian Geographer*, 57. doi: doi:10.1111/cag.12029

Bhopal, K., (2018). *White Privilege: The myth of a post-racial society*, Bristol, Policy Press.

Burman, E. and Miles, S., (2018). Deconstructing supplementary education: From the pedagogy of the supplement to the unsettling of the mainstream, *Educational Review*. doi: doi:10.1080/00131911.2018.1480475

Cahill, C., Alvarez Gutiérrez, L. and Quijada Cerecer, D. A. (2016). A dialectic of dreams and dispossession: The school-to-sweatshop pipeline, *Cultural Geographies*, 23, 121–137.

Clennon, O. D. (2014). *Alternative Education and Community Engagement: Making education a priority*, Basingstoke, Palgrave Macmillan.

Cook, M. (2013). Expatriate parents and supplementary education in Japan: Survival strategy or acculturation strategy, *Asia Pacific Education Review*, 14, 403–417.

Conteh, J. (2007). Culture, languages and learning: Mediating a bilingual approach in complementary Saturday classes. In *Multilingual Learning: Stories from schools and communities in Britain*, J. Conteh, P. Martin, and L. Robertson (eds.), Stoke-on-Trent, Trentham Books, 119–134.

Creese, A. (2009). Building on young people's linguistic and cultural continuity: Complementary schools in the United Kingdom, *Theory into Practice*, 48, 267–273.

Darling, J. (2017). Forced migration and the city: Irregularity, informality, and the politics of presence, *Progress in Human Geography*, 41, 178–198.

Davis, J. L. (2013). *Survival Schools: The American Indian movement and community education in the Twin Cities*, Minneapolis, University of Minnesota Press.

Derrida, J. (1976). *Of Grammatology*, Baltimore, JHU Press.

Dixon-Román, E. (2010). Inheritance and an economy of difference: The importance of supplementary education, *Educating Comprehensively: Varieties of Educational Experiences*, 3, 95–112.

Dove, N. D. E. (1993). The emergence of black supplementary schools resistance to racism in the United Kingdom, *Urban Education*, 27, 430–447.

Dwyer, C., Gilbert, D. and Shah, B. (2013). Faith and suburbia: Secularisation, modernity and the changing geographies of religion in London's suburbs, *Transactions of the Institute of British Geographers*, 38, 403–419.

Evans, D., and Gillan-Thomas, K. (2015). *Descriptive Analysis of Supplementary School Pupils' Characteristics and Attainment in Seven Local Authorities in England*, London, Paul Hamlyn Foundation.

Farrales, M. (2017). Delayed, deferred and dropped out: Geographies of Filipino-Canadian high school students, *Children's Geographies*, 15, 207–223.

Finn, M. (2016). Atmospheres of progress in a data-based school, *Cultural Geographies*, 23, 29–49.

Francis, B., Archer, L. and Mau, A. (2009). Language as capital, or language as identity? Chinese complementary school pupils' perspectives on the purposes and benefits of complementary schools, *British Educational Research Journal*, 35, 519–538.

Gerrard, J. (2011). Gender, community and education: Cultures of resistance in socialist Sunday schools and black supplementary schools, *Gender and Education*, 23, 711–727.

Hamilton, C. V. and Ture, K. (1967). *Black Power: Politics of liberation in America*, New York, Vintage.

Holloway, S. L. and Pimlott-Wilson, H. (2014). Enriching children, institutionalizing childhood? Geographies of play, extracurricular activities, and parenting in England, *Annals of the Association of American Geographers*, 104, 613–627.

Holmwood, J. and O'Toole, T. (2017). *Countering Extremism in British Schools? The truth about the Birmingham trojan horse affair*, Bristol, Policy Press.

Hopkins, P., Olson, E., Pain, R. and Vincett, G. (2011). Mapping inter-generationalities: The formation of youthful religiosities, *Transactions of the Institute of British Geographers*, 36, 314–327.

hooks, b. (1989). Choosing the margin as a space of radical openness, *Framework: The Journal of Cinema and Media*, 36, 15–23.

hooks, b. (2003). *Teaching Community: A pedagogy of hope*, London, Routledge.

Katz, C. (2017). The Angel of Geography: Superman, Tiger Mother, aspiration management, and the child as waste, *Progress in Human Geography*. doi: doi:10.1177/0309132517708844

Kenner, C. and Ruby, M. (2012). Co-constructing bilingual learning: An equal exchange of strategies between complementary and mainstream teachers, *Language and Education*, 26, 517–535.

Kraftl, P. (2013). *Geographies of Alternative Education: Diverse learning spaces for children and young people*, Bristol, Policy Press.

Lipman, P. (2011). Contesting the city: Neoliberal urbanism and the cultural politics of education reform in Chicago, *Discourse: Studies in the cultural politics of education*, 32, 217–234.

Lipman, P. (2013). *The New Political Economy of Urban Education: Neoliberalism, Race, and the Right to the City*, London, Taylor & Francis.

Lytra, V. and Martin, P. W. (2010). *Sites of Multilingualism: Complementary schools in Britain today*, London, Trentham Books Limited.

McCarty, T. L. (2002). *A Place to Be Navajo: Rough Rock and the struggle for self-determination in indigenous schooling*, London, Routledge.

McCreary, T., Basu, R. and Godlewska, A. (2013). Critical geographies of education: Introduction to the special issue, *The Canadian Geographer/Le Géographe canadien*, 57, 255–259.

McKinnon, S., Waitt, G. and Gorman-Murray, A. (2017). The Safe Schools Program and young people's sexed and gendered geographies, *Australian Geographer*, 48, 145–152.

Myers, K. and Grosvenor, I. (2011). Exploring supplementary education: Margins, theories and methods, *History of Education*, 40, 501–520.

Noble, G. (2017). Asian fails' and the problem of bad Korean boys: Multiculturalism and the construction of an educational problem, *Journal of Ethnic and Migration Studies*, 43, 2456–2471.

NRCSE National Resource for Supplementary Education (2015). *Community in Education, presentations at the Community in Education Conference*, London, London Metropolitan University.

Paul Hamlyn Foundation (2015) *Supplementary Schools: Descriptive analysis of supplementary schools pupils' characteristics and attainment in seven local authorities*, London, Paul Hamlyn Foundation.

Pini, B., Gulson, K. N., Kraftl, P. and Dufty-Jones, R. (2017). Critical geographies of education: An introduction, *Geographical Research*, 55, 13–17.

Reay, D. and Mirza, H. S. (1997). Uncovering genealogies of the margins: Black supplementary schooling, *British Journal of Sociology of Education*, 18, 477–499.

Riley, C. and Ettlinger, N. (2011). Interpreting racial formation and multiculturalism in a high school: Towards a constructive deployment of two approaches to critical race theory, *Antipode*, 43, 1250–1280.

Robbins, C. G. (2008). *Expelling Hope: The assault on youth and the militarization of schooling*, New York, SUNY Press.

RSA (2015). *Beyond the School Gates. Developing the Roles and Connections of Supplementary Schools*, London, Royal Society for the Encouragement of Arts, Manufactures and Commerce.

Sellgren, K. (2015). David Cameron: Prime Minister warns over extremist teaching. *BBC*, 7 October. Available at: www.bbc.com/news/education-34464137 [last accessed 21 April 2016].

Shain, F. (2011). *The New Folk Devils: Muslim boys and education in England*, London, Trentham Books Ltd.

Sneddon, R. and Martin, P. (2012). Alternative spaces of learning in East London: Opportunities and challenges, *Diaspora, Indigenous, and Minority Education*, 6, 34–49.

Strand, S. (2007). Surveying the views of pupils attending supplementary schools in England, *Educational Research*, 49, 1–19.

Tereshchenko, A. and Grau Cárdenas, V. V. (2013). Immigration and supplementary ethnic schooling: Ukrainian students in Portugal, *Educational Studies*, 39, 455–467.

Vertovec, S. and Wessendorf, S. (eds) (2010). *Multiculturalism backlash: European discourses, policies and practices*, London, Routledge.

Watkins, M. and Noble, G. (2013). *Disposed to Learn: Schooling, ethnicity and the scholarly habitus*, London, Bloomsbury Publishing.

Wei, L. (2006). Complementary schools, past, present and future, *Language and Education*, 20, 76–83.

Wilson, H. F. (2013). Collective life: Parents, playground encounters and the multicultural city, *Social & Cultural Geography*, 14, 625–648.

Wilson, H. F. (2015). An urban laboratory for the multicultural nation? *Ethnicities*, 15, 586–604.

Wilson, H. F., and Warren, S. (2019, in press). Critical geographies of education: Race, ethnicity, and supplementarity, *Progress in Human Geography*,

Yuval-Davis, N., Wemyss, G. and Cassidy, K., (2018). Everyday bordering, belonging and the reorientation of British immigration legislation, *Sociology*, 52, 228–244.

Zhou, M. (2008). The ethnic system of supplementary education: Non-profit and for-profit institutions in Los Angeles' Chinese immigrant community. In Shinn, M. and Yoshikaw, H. (eds.) *Toward Positive Youth Development: Transforming schools and community programs*, 229–251. Oxford, Oxford University Press.

7 Not an 'other' childhood

Child labour laws, working children and childhood in Bolivia

Chris Willman

Chapter summary

The high incidence of children that work in Bolivia means that experiences of childhood differ from the normative view of childhood experienced in the global north. Bolivia is a unique place to study working children and child labour given that it has a very active movement of working children and lowered the legal minimum working age to 10 in 2014. This was during a period when other Latin American countries, such as Ecuador, Colombia and Argentina, raised or committed to raising the minimum age. This chapter, based on fieldwork conducted between October 2016 and April 2017, is an exploration of childhood in relation to child labour in the socio-political-legal context of Bolivia. It is a contribution to a new body of literature that looks at working children as opposed to child labour in its worst forms, and draws upon a normal vs. other binary approach to childhood, whereby work has no place in a 'normal' childhood, which should be dedicated to learning and play. This approach is largely based on a middle-class childhood of the global north. The chapter argues that a working childhood in Bolivia is 'normal' within the national context, and would only be considered 'other' if approached from a global north perspective. The chapter finds that Bolivia's new child labour law at once compliments and rejects current international standards. Working children assume responsibility to provide for their families financially, whilst combining work and education. In addition, many children have experienced a very politicised childhood through participation in organised unions and campaigns for the right to work.

Introduction

Child labour is a global phenomenon. UNICEF and the International Labour Organisation (ILO) estimate that 152 million children are engaged in child labour globally, just under 50% of whom work in hazardous conditions (ILO, 2017).[1] A rise in global concern towards child labour was witnessed at the beginning of the twentieth century, and followed with the ratification of the United Nations Convention on the Rights of the Child (UNCRC), which sought to protect children from economic exploitation and work that could be considered dangerous or harmful to their health, development or education. Studies of child labour have been diverse and widespread, for example looking at economic perspectives (see Basu and Van, 1998; Grootaert and Kanbur, 1995), the relationship with schooling and education (see Ravallion and Wodon, 2000; Jensen and Nielsen, 1997), children's experiences (see Woodhead, 1998) and definitional issues (see Bourdillon, 2006; Myers, 1999). More recently, research is starting to look at the positive aspects of children working, as well as the negative, and is making an increased distinction between working children and child labour, to detract from any negative connotations associated with the term 'child labour' (Bourdillon et al., 2010).[2] As Liebel (2004) expresses, since child labour became a social problem to be tackled over a century ago, working children were ignored. Research only concentrated on child labour and children as helpless victims (ibid.). There are many differences between the various types of work that children undertake, but child labour is commonly accepted as any harmful work and, according to the ILO, work that deprives children of their childhood (ILO, 2004). Child work on the other hand can be regarded as positive and can aid development, and may include work done in a home setting or whilst employed in family businesses or by third parties. This chapter is an exploration of childhood in relation to working childhoods in the socio-political-legal context of Bolivia, acknowledging children as social actors, whose work is an expression of their agency.

Bolivia is a unique country within which to examine working children. Often cited as one of the poorest countries in South America with high levels of poverty,[3] not only does it have a high incidence of children that work, around 28% of 5–17 year olds, but it also has a very politically active movement of working children, organised geographically into unions. It is also the only country to explicitly permit children to work from the age of 10 years old, which has made international headlines since the new law was passed in 2014 (Eaton, 2018; Bocking and Paz-Soldan, 2016; McQuade, 2014; Watson, 2014). This was during a period when other Latin American countries, such as Ecuador, Colombia and Argentina, raised or committed to raising the minimum age.

The high incidence of children that work in Bolivia means that the experience of childhood differs from the normative view of childhood experienced in the global north encouraged by the international standards on child labour. Previous child labour approaches have called for the abolition of all types of child labour in order to protect children from the harm that work can cause (Bourdillon et al., 2010). This approach has been criticised however as little thought or attention is given to what happens to these children once they are removed from such situations. Moreover, the global north perspective has failed to recognise the opportunities that can be drawn from children working, and denies working children their agency (ibid.). Methods for combatting child labour are now concentrated on the abolition of the worst forms of child labour and allowing certain types of work, as this chapter will shortly highlight.

The relationship between working children, child labour and childhood is compound, but has often been understood through binaries. Bourdillon et al. (2006) summarise two different approaches. One is that childhood is a time to be free from work and employment, and instead dedicated to learning, play and fun. In this 'normal' approach, childhood is essentially different to adulthood, and any other childhood is considered as 'lost or stolen' (ibid.: p. 1202). This includes a childhood of work. The second, or 'other', approach recognises different material and cultural conditions in which childhood exists, and childhood as 'continuous with the adult world, with children gradually moving into the activities of adults as their competences develop and opportunities arise' (ibid.). Work in this approach is part of childhood development.

This chapter aims to argue that a working childhood in Bolivia, when considered in a national context, is not necessarily an 'other' childhood. Rather, in the Bolivian context (as in the Ghanaian, Tanzanian and Bangladeshi contexts explored by Boampong, Pearson and Afroze in other chapters of this book) a working childhood is very common. This chapter will first examine the background to this research; the socio-political context of Bolivia. It will then explore how and what sort of childhood is encouraged by international and national standards on child labour, by looking specifically at the UNCRC, ILO Conventions 138 and 192, and the Bolivian Code for Children and Adolescents. Lastly, drawing on my own research, the chapter will then discuss how childhood is experienced by working children in Bolivia, concentrating on ideas of 'responsible' childhood and 'political' childhood.

Background

Bolivian context

According to the most recent survey of child labour in Bolivia conducted in 2008, there are approximately 850,000 working children,[4] or just under 28% of 5–17 year olds (ILO and INE, 2008). The majority of these, approximately 800,000, were below the recommended minimum age of 14 as set by the ILO and, at the time the survey was conducted, the Bolivian state. 37.9% attend school alongside work, and 79.6% of working 5–13 year olds are considered to be in dangerous work. There are more child workers engaged in work in rural areas compared to urban (55%–45%), although some believe that the numbers for rural areas could be misrepresented and actually higher than this.[5]

A combination of poverty and migration is thought to encourage child labour in Bolivia (ILO, 2014). During the second half of the twentieth century, 62% of Bolivia's rural population moved to urban areas (INE, 2001). This was encouraged by the belief of 'progress, development and modernity' attached to urban areas (ILO, 2014: p.16). In reality, the urban areas offered low paid, insecure and informal sector employment which meant children often had to work in order to support the family. During the same period of rural to urban migration, the population of under 18 year olds increased by 20% (INE, 2001). Commonly, children are employed in retail and light service jobs that are generally considered safe and appropriate for children. Dangerous work that children undertake includes working in brick kilns, as well as silver and tin mines. In addition to this, in many rural areas, forms of work such as agricultural labour are a way of life for children and young people and can be considered as indigenous 'culture'.[6]

Methods

This chapter is based on data collected during fieldwork in La Paz and El Alto in Bolivia between 2016 and 2017. The study used a mixed-methods approach. Over twenty interviews were conducted with members of government, international and national NGO's, working children's unions and various other experts. Various documents, laws and reports from the UN, ILO, Bolivian government and working children's unions were also analysed using a combination of thematic and content analysis. In addition, 174 sentence completion tasks were conducted with children between the ages of 12 and 17, some of whom worked. The sentence completion tasks were 15 sentences and questions organised on a paper questionnaire that the participants were given the option of completing. For example, some of the sentences and questions relevant to this chapter were, '3. I work because...', '5. Work for me means....', and '7. I think child labour means....'. The participants were recruited through two schools in La Paz, one public and one private, and the data collection occurred during the school day. The method was inspired by Morrow (1999), who used the questionnaire in a worksheet form, and Rafaelli et al. (2001), who conducted it orally.

Laws, standards and childhood

There are four international standards and one national law that apply to children working in Bolivia: the UN Convention on the Rights of the Child (1989); ILO standards 138 and 182 (Minimum Age Convention, 1973, and Worst Forms of Child Labour, 1999); Sustainable Development Goal 8.7; and Código Niña, Niño y Adolescente (Children and Adolescents' Code, or law no. 548). These standards and legislation will be discussed in turn in the section that follows.

The Convention on the Rights of the Child was a landmark international standard. It brought to the fore children in international human rights treaties, where they had previously been ignored (Fontana and Grugel, 2015; Grugel and Peruzzotti, 2007). The convention pledges basic human rights for children: the right to a childhood, to be educated and healthy, to be treated fairly and to be heard (UN General Assembly, 1989). It references child labour and briefly states that governments should protect children from work that is dangerous or might be a detriment to their health or education. It does not ban work completely however, and allows for children to help at home in safe and age-appropriate ways. Any work children do should not interfere with any of their other rights, including the right to education, and the right to relaxation and play. ILO Convention 138 sets the minimum age that a person may participate in economic activity at 15 (ILO, 1973). Special case countries may set the minimum age at 14, with the aim of raising it to 15. This was the case with Bolivia, when they ratified 138 in 1997. ILO Convention 182 lists the worst forms of child labour, including modern slavery, forced recruitment to the military or armed conflict, sex work, illicit work and any other work that is likely to harm the health, safety and morals of children (ILO, 1999). Bolivia ratified this convention in 2003. Finally, Sustainable Development Goal 8.7 demands the prohibition and elimination of the worst forms of child labour, and by 2025 an end to child labour in all its forms.

Nationally, the Children and Adolescents Code was ratified on 17 June 2014 and is the first law in relation to child labour in the world to have been created with involvement from children (Liebel, 2015). The code covers all aspects of children's lives, from schooling and health to work and family, and is done in their best interests (Estado Plurinacional de Bolivia, 2014). The law not only sets out their rights, but also their duties (ibid.). The role of enforcing this code is down to an ombudsmen service, *La Defensoria de la Niñez y Adolescencia.* [7] The key points pertinent to this chapter are that:

- Children between 10 and 12 can only work as self-employed or for family members, whereas over 12s can work for a third-party employer.
- Those working between the ages of 10 and 14 must seek approval from the *Defensoría*, with approval contingent on a medical examination.[8]
- Children over 14 must receive an equivalent minimum monthly wage (which could be paid daily, weekly or monthly), no less than that earned by adults (the national minimum is currently around US$250 per month). From the interviews conducted however, this is not strictly followed. The law does not mention a minimum wage for under 14s.
- Under 14s cannot work more than 6 hours a day, and 30 hours a week. Over 14s must not work more than eight hours, or 40 hours per week. Working days must finish before 10pm.
- Employees are entitled to two hours paid study time.[9]
- All children, working or not, must still attend school.
- The law prohibits children from working in industries which may affect health, such as in sugar cane plantations, mining, brick factories, sale of alcoholic beverages, and rubbish collection. Other work that is illegal includes anything of a sexual nature, or that is dangerous, unhealthy or threatening to the dignity and integrity of children.
- Work in a family or community setting is permitted at any age, as a recognition of the aforementioned indigenous culture that is practiced in many rural areas.

The new law caused international outcry being implemented against the wishes of many international organisations, including the ILO, UNICEF and Human Rights Watch, and was derided by international media (Howard, 2014; McQuade, 2014).

Each of the standards' definitions of childhood shows evidence of differences in the expectations of childhood. The UNCRC defines children as anyone under the age of 18, unless that country specifies that majority is obtained earlier. This in itself is problematic as experiences, needs and interests change and develop throughout childhood. The Bolivian state, in the new code, distinguishes between children and adolescents, a child being under 12 and an adolescent as 12 and over. From the fieldwork conducted for this study between 2016 and 2017, the children who participated in the sentence completion task were asked to self-identify themselves as a child, an adolescent, a young person (*joven* in Spanish, a colloquial term for young person) or an adult. Very few of the 12–13 year old participants self-identified as a child (5 out of 174). The majority of the 12 and 13 year olds identified themselves as *jovenes*. Those over the age of 14 self-identified nearly 50/50 as *jovenes* and adolescents. None identified as an adult, not even the one participant at 18 who was legally classed as such.

This more specific age distinction, between children, adolescents and *jovenes*, can help diminish any connotations afforded to the notion of children working, especially when using a blanket definition of children as under 18. Work can be considered more acceptable as an adolescent or a young person at 14, rather than say as a child at 14. Although the age of the person working does not differ, the label attached to it can change perceptions. While the idea of a childhood consisting of work can sit uneasily with one approach to childhood, the idea of working during adolescence is harder to disagree with, across any approach to childhood, 'normal' or 'other'. Furthermore, it supports the agency of children better by allowing them to self-identify, which the UNCRC does, calling for consultation of children on their views where necessary.

It was highlighted during various interviews with members of working children's unions and the ministry of labour that the new Bolivian code refers to working children and not child labour. Manfred Liebel (2015) also points out that it refers to protection *at* work as opposed to protection *from* work, which has previously been the norm in standards on child labour. In terms of childhood this means that there is an attempt by the law to distance itself from the negative connotations of child labour, which usher in perceptions of what is now referred to and mentioned before as the worst forms of child labour: harmful and sometimes forced work. As one of the interviewees, a child psychologist, explained 'the term *labour* suggests exploitation, but *work* is different.' In fact, the work that many children and adolescents do in Bolivia mirrors much of the work that children and adolescents in the global north undertake. Common jobs that the participants of this study undertook involved working in restaurants, local shops or helping in family businesses such as at market stalls. Therefore, the law protects children and adolescents *in* these occupations with minimum wages and maximum work hours, rather than protecting children *from* them. In doing so, it also supports the agency of children and recognises them as social actors, by not defining them as something to be protected in the context of work; rather, accepting and respecting their decision to work.

The code however is not in complete deviance to the international standards. It has in common with the international standards the suggestion that childhood should be devoid of any harmful work. This is done with the aim of preventing any detriment to young people's health or morality. Also in common across national and international codes and standards is the placement of education as of high importance for children. Generally, the international standards concentrate on formal schooling as education and the most important component of a normal and ideal childhood.

The new code in Bolivia, then, at once compliments and rejects current international standards. These are arguably standards of a middle class and minority world childhood whereby childhood is seen as a time to be dedicated to education, play and leisure, not work. The new code supports and rejects this; education should be of a primary concern, but work is acceptable too. By proposing that partaking in work, as well as formal schooling, is part of childhood development, the code crosses the normal/other binary on childhood, which have previously suggested that work and education are not considered compatible within childhood. The code legitimises children working below the age of 15 by giving the right to work at the age of 10. By giving them the right to work, and recognising the value and importance of children's work, the code strikes a new position amongst historic and contemporary global child labour policies. In so doing, the code demonstrates a perceived value not only of children's work, but also of the children who work, and therefore bestows upon child workers not only responsibility, but also legal rights and protection.

Responsible childhood

Following from the notions of childhood encouraged by international and national standards addressing child labour, the next point of discussion is around the idea of a 'responsible childhood' (the concept of intergenerational contracts of responsibility is also discussed by Pearson, this volume, in relation to street children). One way of understanding working childhoods in more depth is through the concept of responsibility, which emerges through analysis of the Bolivian Children and Adolescents code and the sentence completion task responses. As aforementioned, the new code sets out the duties of children and adolescents in Bolivia, although these are not a prerequisite for their other rights (Estado Plurinacional de Bolivia, 2014). Article 158 states they are to 'assume their responsibility as active subjects in the construction of the society' and 'act with honesty and co-responsibility[10] in all circumstances' (ibid.: p.86). Liebel (2015) also points out that family and community work as mentioned in the code is allowed because it permits their formulation as active and responsible citizens. The UNCRC focuses solely on the responsibility of parents or guardians for children, showing that they are viewed as something to be responsible over. This section, on the other hand, shows that children exercise agency by being responsible.

The notion of responsibility appeared frequently in the sentence completion task responses from the participants, particularly to the questions 'I work because...', 'I like work because...', 'Work for me means...' and 'What does it mean to work at your age?' These responses mentioned responsibility directly, such as 'Work means to be responsible in life', 'Work for me means responsibility' and 'Work for children and adolescents should not be illegal because you learn to mature and be responsible'. But the responses also had a sense of responsibility in them, such as 'I work to help my family' and 'I work to help my parents with bills', amongst other similar responses.

So, what does all this mention of responsibility mean in terms of childhood? The responses from children and adolescents themselves, as well as the duty of responsibility in the new code, suggest three things. Firstly, that Bolivian children and adolescents are *taking on* a responsibility, secondly that they *feel* responsible for their families, and finally that this responsibility is also something *learned, gained and developed* through work. Taking on a responsibility means helping to provide for family, but also helping in tasks and duties at home, such as care roles. This is an economic contribution in terms of wages earnt from work, but also a contribution of time at family run businesses, such as restaurants or small street shops, as well as looking after other family members such as younger siblings. Feeling responsible becomes a motivation and justification for taking on responsibility. Responses suggest that the participants felt responsible for both their parents and siblings and expressed a need to help them. They felt that they should help their families, as it is their responsibility to bear, reflecting the same social concept of intergenerational contracts as that discussed by Pearson (this volume) in the Tanzanian context. In contrast to the feelings of young people, the Bolivian code encourages responsibility towards state and society through its list of duties for children and adolescents, rather than to families (Estado Plurinacional de Bolivia, 2014).

Respondents also clearly positioned responsibility as a skill or value, to be learned, gained and developed through work. One specifically mentioned that work for children and adolescents should not be illegal because 'you learn to value and become responsible'. UNATsBO, the national working children's union in Bolivia, also cites responsibility as a key component of working as a child and adolescent. In their publication 'Mi fortaleza es mi trabajo' (My strength is my work) UNATsBO argue that work allows children and adolescents capacity to become more responsible and gain responsibilities (UNATsBO, 2010).[11] By contrast, the international or Westernised view of childhood, in discouraging work, suggests that children should be absolved of

such associated responsibility. In addition, responsibility is often a trait associated more with adulthood than childhood. This research however shows
that it is prevalent in the childhoods of Bolivian children and adolescents;
they take it on themselves and are encouraged to do so. Thus, childhood in
Bolivia may not be viewed as entirely distinct from adulthood in the way that
international charters might present it.

From this, the question arises as to whether work teaches children and
adolescents to become responsible, or whether children and adolescents who
have a sense of responsibility decide to work. Arguments can be put forward
for both answers. Based on the responses from this study, it seems clear that
the participants here believe that work teaches responsibility and thus it is fair
to consider work in itself as a type of education. However, that is not to dismiss that some children who feel responsible work for their family, and this
does not have to be mutually exclusive with the previous point. Nevertheless,
the more sceptical participants suggest that the reality is that situations of
poverty mean that children must work, and that the idea of responsibility
may be a nicer, more positive narrative that they give for their situation.
Regardless of this, in expressing their justification of responsibility, children
and adolescents are exercising their agency.

Political childhood

An alternative way of examining working childhoods in Bolivia is through the
concept of political childhood. As mentioned previously, the movement of
working children is a very active part of civil society in Bolivia. Collectively
known as NATs (niños, niñas y adolescentes trabajadores / children and
adolescent workers) working children are organised into unions with backing
from international NGOs, such as Save the Children and Terre des Hommes,
as well as backing from MOLACNATs, the movement for Latin America and
the Carribbean. UNATsBO and ASONATs are the largest national groups in
Bolivia, with smaller sub-organisations located in the major cities. There are
approximately 150,000 members in NATs across Bolivia, with members
commonly joining at age 12, though some are as young as 7 or 8.

During interviews in the field, members of NATs were quick to point out
the benefits of being members of the unions. A lot of work that children do,
for example shoe shining and vending on the streets, can be lonely and isolating. Membership however has given them a safe space in which to meet,
organise and discuss their experiences, as well as a social space to meet others
in similar situations. Being a member allows them to develop skills such as
teamwork, leadership, public speaking and debating – vitally important to
those who didn't partake in mainstream schooling and education. One participant, David, stated how '...we have been empowered because we have been
able to meet and talk with senators, congressmen, with the vice president and
chancellors.' A NATs collaborator talked proudly about being able to take a
young member to Geneva to the ILO's head office. Members also developed

political awareness through exposure to political processes, spaces and institutions. From a young age those members are exposed to political mechanisms and structures, by partaking in meetings with the government, holding meetings themselves as well as reading and drafting publications.

The adults who work to help run the working children unions see these young members as the future politicians of Bolivia. It is common that when members reach 18 and are no longer allowed to be official members due to their newly acquired status as adults, they often stay on as collaborators to advise and steer members. Indeed, Evo Morales, Bolivian President at the time the code was passed, is a former child worker, having worked on his parents' farm. He is quoted as saying that to eliminate work for children in Bolivia would be to eliminate people's social conscience (Simpson, 2014).

The involvement of working children in drafting the new child labour policy is unique globally. Before the adoption of the new code in 2014, NATs also drafted their own code which suggested no lower limit to a legal working age (UNATsBO, 2010; Liebel, 2015).[12] The Bolivian state then released their own draft law. When it was rumoured that work would still be banned for those under 14 years, working children took to the streets in December 2013, which included violent clashes with the police in La Paz, widely reported by the media (Liebel, 2015). Following these protests over Bolivia's draft law, the wishes of working children were taken into consideration and representatives from UNATsBO were invited to a series of meetings with the government to express their opinions and help formulate the new code. In addition to political participation through working unions, the new code also allows children and adolescents in Bolivia to participate politically through Children's and Adolescents' committees. The committees' role is to be involved in formulating, implementing and monitoring policy relevant to children at different levels, from the municipal right up to the national. However, the committees' capacity is arguably limited through lack of funding.

Prior to the new code, which granted children the right to work from 10 years old, members were frustrated by their lack of rights and citizenship. One interview with a current NATs collaborator and former member led to him stating that they were frustrated because 'I have so many rights but that they will not activate until I turn 18 ¡No! It is ridiculous that you have so many rights and say, until you are 18 years old, you are not a citizen'. Now children have the right to work, many see themselves recognised by state and society and participate in political processes. There is, however, one major limitation to political participation rights, and that is that citizens in Bolivia cannot vote until they are 18.

Exposure to political processes, spaces and structures, which has also involved being tear-gassed and beaten at the hands of the Bolivian police during the aforementioned protests in December 2013, can be considered extremely deviant from the expectancies and experiences of a 'normal' childhood in an international context. The ideal vision of a protected childhood encouraged by the international standards would not include a childhood spent actively campaigning for the right to work, or involved in violent cla-shes with national police in the streets, which is the case in Bolivia. In some ways, this childhood experience has more in common with adulthood, and has bestowed upon children both positive and negative aspects of work; chil-dren now have the right to work and participate in political processes, but they had to fight for it and suffered violence.

Conclusion

A working childhood in Bolivia cannot necessarily be considered as an 'other' childhood. It is only othered when looking at it from the perspective of the global north that promotes a 'normal' childhood as a time to be free from work and children as needing to be protected from work. Rather, a working childhood in Bolivia is very common. It is a childhood of the majority world that respects, encourages and grants children the right to work. Yet, just as there may not be a 'normal' typology for children globally, in the global north or south, there is not one typology for children in Bolivia. Rather, childhoods in Bolivia, as they are in many other contexts across many spectrums, are unique, messy and complex to understand, as other chapters in this book will attest. This chapter has sought in part to understand a cross section of chil-dren and adolescents in Bolivia, those who work, and their childhoods by concentrating on the idea of a political childhood and the notion of respon-sibility. This is helped further by exploring what childhood is encouraged by national and international standards in relation to work. There is also the voice of those children who do not work, and their childhood that has yet to be explored and understood. In addition, it was presented that the concept of adolescence could be explored further as a way to bridge the gap between these social constructions of adulthood and childhood.

The national standards of Bolivia both reject and compliment interna-tional standards and their ideals of childhood (to be free from work and dedicated to play and learning). Education is of upmost importance in a child's life, but work can be important too. Taking on, feeling and develop-ing responsibility is a key element of working childhood in this context, as is exposure to political processes, institutions and spaces and the claiming of rights. These elements in many ways link childhood and adulthood. Par-taking in politics (which Hall and Pottinger also discuss in their chapter, this collection) and being responsible are usually reserved for adulthood in many contexts, but in this instance the idea of political childhood and responsible childhood link both adulthood and childhood. Gaining a better under-standing of childhoods, and in this case working childhoods, is crucial for not just academic debate but also policy on child labour and working chil-dren. Furthermore, by allowing children and adolescents an input in this, which the fieldwork that has informed this chapter aimed to do, gives agency to children and allows them to express their opinions, and us to understand their experiences.

Notes

1 This is for children between the ages of 5 and 17. Briefly, hazardous work is descri-bed as directly endangering the health, safety and moral development of children.
2 This is not an exhaustive list of the literature on child labour, and there are many more publications one could consult.
3 This of course depends on how poverty is measured. For more information see World Bank, 2018.
4 Where 'working children' are defined as children engaged in economic activity for at least one hour per week.
5 This is due to the complexities of conducting surveys in rural areas, and because this work predominately occurs in a family and community setting without remu-neration, thus can be obscured more easily. While this sort of work, mostly agri-cultural, may be considered work by global north standards, in many indigenous communities it is considered more of a duty.
6 For more see Chambi Mayta, 2017.
7 The enforcement and management of the law has come under much criticism, due to a lack of institutional capacity related to funding. For more information see Liebel, 2015 and Fontana and Grugel, 2016.
8 However, interviews showed that this process is unclear and unbeknown to many.
9 Similarly to the footnote above, how this is to be put into practice is unknown.
10 This is *corresponsabilidad* in Spanish which translates to co-responsibility or joint responsibility.
11 This document was published to recognise the demands and needs of working children, as well as present UNATsBO's draft law for working children.
12 This was entitled 'proposal for the recognition, promotion, protection and defence of the rights of children and adolescent workers' and was written at the end of their 'Mi Fortaleza' document.

References

Basu, K. and Van, P.H. (1998). 'The economics of child labor', *The American Economic Review*, 88(3), pp. 412–427.

Bocking, D. and Paz-Soldan, Y. (2016). 'Child labor in Bolivia is legally permissable'. *Spiegel*. Available at: www.spiegel.de/international/tomorrow/child-labor-in-bolivia-is-legally-permissable-a-1130131.html (Accessed: 23 April 2018).

Bourdillon, M., *et al.* (2006). 'Children and work: A review of current literature and debates', *Development and Change*, 37(6), pp.1201–1226.

Bourdillon, M.*et al.* (2010). *Rights and wrongs of children's work*. New Brunswick, NJ: Rutgers University Press.

Chambi Mayta, R.D. (2017). 'Living Well, child labor, and indigenous rights in Bolivia'. *Latin American and Caribbean Ethnic Studies* , 12(2), pp.95–112.

Eaton, T. (2018). Bolivia child labor law allows world's lowest minimum age. *USA Today*. Available at: www.usatoday.com/story/news/world/2018/01/09/bolivia-lets-10-year-old-kids-work-under-worlds-youngest-child-labor-laws/1011101001/ (Accessed: 23 April 2018).

Estado Plurinacional de Bolivia. (2014). *Ley No 548 Codigo Niña, Niño y Adolescente*. La Paz, Ministerio de Justicia.

Fontana, L.B. and Grugel, J. (2015). 'To eradicate or to legalize? Child labor debates and ILO Convention 182 in Bolivia', *Global Governance: A Review of Multilateralism and International Organizations*, 21(1), pp.61–78.

Fontana, L.B. and Grugel, J. (2016). '¿Un nuevo rumbo para el trabajo infantil en Bolivia? Debates y polémicas sobre el Código de la Niñez', *Nueva Sociedad*, 264, pp. 87–98.

Grootaert, C. and Kanbur, R. (1995). 'Child labour: An economic perspective', *International Labour Review*, 134, p. 187.

Grugel, J. and Peruzzotti, E. (2007). 'Claiming rights under global governance: Children's rights in Argentina', *Global Governance: A Review of Multilateralism and International Organizations*, 13(2), pp.199–216.

Howard, N. (2014). On Bolivia's new child labour law. *Open Democracy*, 6 November. Available at: www.opendemocracy.net/beyondslavery/neil-howard/on-bolivia%e2%80%99s-new-child-labour-law (Accessed: 15 February 2019).

ILO (International Labour Organisation). (1973). *Convention C138: Minimum age convention 1973*. Geneva, ILO.

ILO (International Labour Organisation). (1999). *Convention C182: Worst forms of child labour convention, 1999*. Geneva, ILO.

ILO (International Labour Organisation). (2004). *Child labour: A textbook for university students*. Geneva, ILO.

ILO (International Labour Organisation). (2014). *Estudio sobre trabajo doméstico de niños, niñas y adolescentes en hogares de terceros en Bolivia*. La Paz: ILO, UNICEF, FENATRAHOB, UMSA, MTEPS.

ILO (International Labour Organisation). (2017). *Global estimates of child labour: Results and trends, 2012–2016*. Geneva, ILO.

ILO (International Labour Organisation) and INE (Instituto Nacional de Estadística). (2008). *Bolivia: Encuesta de trabajo infantil 2008*. La Paz: UNICEF, OIT, Ministerio de Trabajo Empleo y Previsión Social.

INE (Instituto Nacional de Estadística). (2001). *Bolivia: Caracteristica de la Vivienda*. La Paz, INE.

Jensen, P. and Nielsen, H.S. (1997). 'Child labour or school attendance? Evidence from Zambia', *Journal of Population Economics*, 10(4), pp. 407–424.

Liebel, M. (2004). *A will of their own: Cross-cultural perspectives on working children*. London, Zed Books.

Liebel, M. (2015). 'Protecting the rights of working children instead of banning child labour', *The International Journal of Children's Rights*, 23(3), pp. 529–547.

McQuade, A. (2014). 'Bolivia's child labour law shames us all', *The Guardian*. Available at: www.theguardian.com/global-development/poverty-matters/2014/jul/25/bolivia-child-labour-law-exploitation-slavery. (Accessed: 23 April 2018).

Morrow, V. (1999). '"It's cool, … 'cos you can't give us detentions and things, can you?!": Reflections on researching children'. In: P. Milner and B. Carolin (eds.), *Time to Listen to Children*, pp. 203–215. London, Routledge.

Myers, W.E. (1999). 'Considering child labour: Changing terms, issues and actors at the international level', *Childhood*, 6(1), pp. 13–26.

Rafaelli, M. et al. (2001). 'How do Brazilian street youth experience "the street"? Analysis of a sentence completion task', *Childhood*, 8(3), pp. 396–415.

Ravallion, M. and Wodon, Q. (2000). 'Does child labour displace schooling? Evidence on behavioural responses to an enrollment subsidy', *The Economic Journal*, 110 (462), pp. 158–175.

Simpson, J. (2014). 'Bolivia becomes first nation to legalise child labour', *The Independent*. Available at: www.independent.co.uk/news/world/americas/bolivia-becomes-first-nation-to-legalise-child-labour-9616682.html (Accessed: 23 April 2018).

UNATsBO. (2010). '*Mi fortaleza es mi trabajo*' *De las demandas a la propuesta. Niños, niñas y adolescentes trabajadores y la regulación del trabajo infantil y adolescente en Bolivia.* Available at: https://tdhsbolivia.org/pdfs/Mi_fortaleza_es_mi_trabajo_Doc_Final.pdf (Accessed: 29 June 2018).

UN General Assembly. (1989). *Convention on the Rights of the Child.* Available at: www.ohchr.org/en/professionalinterest/pages/crc.aspx (Accessed: 15 February 2019).

Watson, K. (2014). 'Child labour laws: A step back for advancing Bolivia?' *BBC News.* Available at: www.bbc.co.uk/news/business-30117126 (Accessed: 23 April 2018).

Woodhead, M. (1998). *Children's perspectives on their working lives: A participatory study in Bangladesh, Ethiopia, The Philippines, Guatemala, El Salvador and Nicaragua.* Sweden: Rädda Barnen.

World Bank. (2018). *Bolivia overview.* Available at: www.worldbank.org/en/country/bolivia/overview (Accessed: 29 June 2018).

8 Unschooling and the simultaneous development and mitigation of 'otherness' amongst home-schooling families

Nadia von Benzon

Chapter summary

This chapter draws on research addressing publicly available blogs published by mothers who home educate their children using the 'unschooling' approach. Unschooling is a form of home education that is child-centred, requiring the child to lead their own learning according to their interests, whilst a parent facilitates. Children who are unschooled can be considered 'other' within a neoliberal context in which formal education is seen as a building block for a healthy economy. Unschooling contradicts the key tenets of education in Western democratic states where value is placed on a homogenised approach to education through a deeply hierarchical system that produces a programme that is designed, administered and assessed by expert adults. This chapter considers the ways in which 'otherness' is recognised in unschooling families, and the techniques that are employed by these families to ensure that this arguably self-imposed 'otherness' is a positive experience for their unschooled children. The blogs demonstrate a variety of techniques, such as the development of online and real-life communities, used by home-schooling mothers. These techniques might be said at once to be attempts to develop a strong identity that embraces otherness and alterity, whilst also finding ways to minimise potential negative impacts of otherness that might hinder their children's opportunities for full engagement in a local community and, later, in the wider economy.

Introduction

Autonomous education, known as unschooling, is a form of home education that is based on an educational philosophy in which children are viewed as competent agents in their own learning (Holt 1995). Unschooling posits that children are naturally curious, and left to their own devices, will seek out knowledge. This approach is lauded by proponents as a type of education that fosters children's innate curiosity about the world, sparking a desire to learn and a life-long love of learning (de Wit, Eagles, and Regeer 2017). Unschooling parents may 'strew', that is, provide access to a variety of resources they deem appropriate or interesting and likely to stimulate their child's educational explorations. However, unschooling children are offered free reign to take up these resources as they please (Griffith 1998). Unschooling does not require that children have access to a single educator or syllabus, but rather that providers facilitate opportunities for exploring knowledge and following up ideas. As such, the philosophy of unschooling exists in opposition to mainstream educational practices that seek to control children's bodies and direct their knowledge and skill formation to meet specific criteria (Barker et al. 2010; Catling 2005). Unschooling becomes the antithesis of mainstream education, particularly in terms of the heightened agency of the child and the democratic nature of decision-making. Moreover, where mainstream education is structured and regulated, unschooling is fluid and dynamic. Operating at the smallest scale of individuals and family units, and where parents remain responsible for the education and welfare of their own children, unschooling offers a level of agency for the child that goes beyond the possibilities of other alternative forms of education. Unschooling might be framed more broadly as an alternative lifestyle that contravenes broadly accepted relationships between families and the state.

Unschooling remains a minority form of home education. However, many home-educating families borrow from unschooling, and certainly it is common amongst home-educating families to attempt to tailor education to each child's specific interests (Liberto 2016). There are likely to be many reasons why unschooling remains relatively unpopular amongst approaches to home education. This may be due to a lack of trust in the underpinning philosophy, and fear that children who are not steered in their learning simply will not learn. It is likely also due to a desire to develop young people who are able to engage in the workforce as adults, or even to access mainstream education at a later date. Therefore, many home-schooling families will recognise a need to ensure their child is literate and numerate, and perhaps is even 'on a par' in terms of national curriculum knowledge with their peers, regardless of their particular interests. There is also some evidence to suggest that unstructured homeschoolers do not have as successful outcomes as those following structured curriculums (Campbell 2012). Given the potential objections to unschooling, it is possible that parents who follow this educational path do so through an explicit desire to avoid conformity and in order to raise children differently and / or to raise children to *be* different. Unschooling might be seen as a deliberate alterity in social and economic engagement, as it minimises the potential for 'mainstream' or state-sanctioned influence on the intellectual, social and emotional development of the child (Collom 2005).

Given not only its lack of institutionalisation, but also its overarching and intentional lack of structure, the daily lived experience of unschooled children, as with home-educated children more generally, will contrast to the experience of their mainstream-schooled peers. Home-educated children experience space and time in a way that contrasts to their mainstream-schooled peers, and are likely to have a different set of networks and relationships within their community and beyond, and a different identity marked by different hierarchies of belonging. Mothers who home educate are also likely to experience time and space differently from other mothers (Lois 2010), given for example, that they will not have periods of the day where their children are in the care of others and where they can access their local communities independently. Nor are home-educating mothers tied to diurnal rhythms imposed by school timetables or journeying between institutions and the home.

Discussion of the experience of unschooled young people contributes to a growing literature developing in the overlap between children's geographies and geographies of education, addressing alternative forms of education and parental, particularly mothers', involvement in education (Kraftl 2013b; Holloway and Pimlott-Wilson 2013; Kraftl 2015). Geographers' interest in otherness around education, which manifests through discussions of alternative education (Kraftl 2013a, 2014), as well as alterity amongst pupils in mainstream schools (Wilson 2014; Holt, Bowlby, and Lea 2017), reflects an interest that lies at the nexus of situated learning and the importance of embodied inhabitation of spaces and places as part of the learning experience. Beyond the immediate physicality of the affect of education are broader questions about the effect of education on young people and society more generally. Additionally, of importance is the role of identity in shaping educational experiences, and vice versa, the role of educational experiences in shaping identity, aspirations and life chances (Holloway et al. 2010; Pini et al. 2017).

The focus of this chapter – exploring a particularly 'other' form of alternative education – offers insight into the experience of families choosing to live and learn in a way that sets them apart from mainstream society in the UK, and in the minority world more broadly. The chapter demonstrates that alterity can be a choice, that doing things differently may lead to isolation that is either chosen or at least accepted as necessary collateral damage within a broader and more important goal of educating differently. However, this chapter also explores the social collaboration and coagulation that can occur amongst those parenting 'otherwise' in order to facilitate social integration from a potentially isolating activity – thus demonstrating potential agency in reducing 'otherness'. This chapter explores the experience of home-educated children in neoliberal Western democracies – here specifically the UK, the USA and Australia. The chapter considers the ways in which being educated at home, particularly following a form of unstructured and child-led home education, leads to a specific childhood geography that is distinct from that of their mainstream-educated peers. The paper reflects a mother's view of unschooling in that it draws on data from individual unschooling blogs authored by home-educating mothers, whilst also written by an author mainstream-schooling her own children.

Home educating, alterity and isolation

Once positioned as a highly subversive, and potentially dangerous, activity, home education is growing in popularity in the UK and many other Western democracies (Gaither 2017; Jolly, Matthews, and Nester 2013; Gaither 2009). The regulation and legalities of home education vary widely from one country to the next (Cooper and Sureau 2007). In Germany, home education is illegal, whilst in the UK, home-educating families do not even need to register.[1] In the US and in Australia, policy addressing home education is within individual state jurisdiction rather than the federal government, so rules vary from state to state. Home education has traditionally been underpinned by a variety of philosophies aligned by Morrison (2014) with those of conscientious objectors. That is to say that, given the ubiquitous uptake of mainstream education, home education has been widely viewed as a protest or objection to this norm. Collom and Mitchell (2005) describe home education as a social movement, with home educators typically motivated either by religious practice or by liberal values – a dichotomy jocularly referred to as the Heaven or Earth approaches (Collom 2005). However, in contemporary society, myriad reasons for home education span from structural, religious or political objections, to specific personal concerns about the ability of mainstream schooling to meet the needs of an individual child (Jolly, Matthews, and Nester 2013; Schafer and Khan 2017). In fact, the numbers of home-educating families, the reasons for doing so, and the approaches to it, have grown so considerably in recent decades that home education is no longer considered a radical approach to education (Gaither 2009; Jolly and Matthews 2017). Home education takes a wide variety of forms, from school at home, in which education is rigidly timetabled and highly structured around clear curriculum, to autonomous education (Gaither 2017). It is the latter practice that forms the focus of this chapter.

Given the explicit alterity of the practice, there is the potential for unschooling families to become isolated (Tollefson 2007). The notion of isolation has long been a key concern in discourse surrounding home education for a variety of reasons, particularly in terms of the risk of parents using home education purposively to isolate children for either cultural and religious reasons – in order to ensure that the children's education reflects the family's priorities and morality (Kunzman 2017; Sherfinski and Chesanko 2016) – or in order to perpetuate abuse (Cooper and Sureau 2007; Webster 2013). However, there is also suggestion that people will home educate as a result of preexisting social isolation (Levy 2009), or where they feel that an identity marker such as race will disadvantage their child in mainstream education (Ray 2015, see also Wilson et al., this volume). Indeed Collom and Mitchell's (2005) paper suggests that families choosing to unschool may do so because their identity is already different to that of the people around them, and they do not wish their child to be educated into the wider cultural norms of the community in which they are resident. Whilst some parents may enjoy the family-focus that social segregation through home education can afford, others find ways, both online and offline, to build social networks to counteract the potential isolation (Jolly and Matthews 2017).

The blogs posts that will become the focus of the remainder of this chapter are all authored by women. This was not an explicit choice in the sampling, but rather a fact that arose from the blogs that were collected. However, the authorship of these blogs reflects the wider literature on gender and home schooling which demonstrates that, by and large, mothers are responsible for providing the education and child care in home-educating families (Lois 2006, 2017; Kraftl 2016). Unsurprisingly, this reflects broader gender disparity in responsibility for other educational activities in mainstream-educating families, such as supervising homework, supervising the school commute, and facilitating extracurricular activities (Holloway and Pimlott-Wilson 2013; Barker 2011; O'Brien 2007, 2005). Given that the data will be drawn from the voices of women, this paper will go on to explore the experiences of home-schooling families as told by the mothers.

Researching unschooling blogs

The initial research design intended to involve the interrogation of child-authored blogs addressing unschooling. Brief exploration of home-schooling blogs showed that there were very few publicly accessible blogs authored by children on this topic. Children's blogs were secure and only accessible to invited readers. It may well be possible that far more of these blogs are written and become apparent to members of the unschooling community, or an established reader of one of these blogs. However, using publicly available online data was a method I was keen to explore in this research in order to reflect on the ethical ambiguity of conducting 'covert' research in public online space (von Benzon 2019). Pragmatically, the research then shifted focus to explore blogs authored by the adults who facilitate unschooling.

Unschooling blogs are not as plentiful as mainstream home-educating blogs, since unschoolers form a small minority of the home education community. There are also few public listings of unschooling blogs. However, home-education blogs are fairly well linked up – whether formerly through a blog carousel[2] such as the 'Homeschooling Linky' or informally through recommended blog lists on individuals' blogs. As such, blogs were found, rather than chosen, and accessed via a snowballing method. In total 14 blogs were analysed, using a grounded theory approach, through NVivo. These blogs were entirely UK, US or Australian, reflecting the patterns in the network and interconnections of English-language bloggers and blog-readers, I suspect, more than the overall geographical dispersion of unschooling. However, the latter is a possibility.

The use of publicly uploaded information to the internet as a data source is an ethical minefield, and one I explore in more detail elsewhere (von Benzon 2019). However, there are perhaps two key ethical challenges unique to working with content posted to blogs and other web hosts such as social network sites: considering intended audience, and status of the author. Snee (2012) argues that the two primary ethical issues around the use of blogs as research data concern whether they can be considered public or private content, and whether those who publish them can be considered subjects or authors. Interestingly, recent developments in English law suggest that as a society we now consider posts to the internet, whether entirely public or within our own semi-open networks, such as our Facebook profile pages, to be public. Moreover, the law considers publishers of this material to have complete agency in the process (von Benzon 2019). With this particular topic, and with this particular group of authors – home-educating mothers writing to connect with the public and each other, to educate, inform and entertain – I was happy to consider the content as 'public' and the writers as competent authors. From this perspective, I mined blogs for qualitative data and approached the analysis and dissemination of research in the same manner as I might have approached thematic analysis of national magazine or news-paper content.

As a children's geographer, I keenly felt the absence of the children's own voices using this research method. The children were present through photo-graphs and anecdotes and through images of the produce of their labour, whether artworks or writing or food. However, all of the blogs were parent-authored and there was no suggestion in any of the 14 blogs that children had directly contributed to their writing or design. Rather, this chapter fore-grounds the perspectives and experiences of mothers. What this offers the collected edition is some insight into the lived experience of one of the key figures in many children's lives. It is not my intention that the mothers' views are used a proxy for the children's experience, but rather that this chapter serves to provide insight into the way in which children's lives might be determined as 'alternative' through distinct choices made by their parents (see also, for example, Santariano, this volume), or through choices made in col-laboration with parents. Throughout the chapter I have used text from the blogs to illustrate the points I have made. The exact wording has been altered to maintain the confidentiality of bloggers whilst providing some reflection of personal experience. Given that extracts have been chosen to represent gen-eralised points, and emanate from a range of different blogs, I have presented the blog text within this chapter without reference to the individual blogs from which they were taken.

Unschooling childhood as 'other'

The families foregrounded in the research largely reflect the demographics of 'typical' (lower) middle class families in the UK, US and Australia (according to a UK-centric understanding of class). Narratives and photographs within the blogs depicted families that were almost exclusively white. Most were two-parent households, although there were two single-parent households within the sample. Most families were described as having one professional parent – one family had two professional parents. Each family's story suggested they lived above the poverty line, and parents were literate. In many ways the demographics of the blogging mothers appeared to reflect those identified in broader discussions exploring mommy blogs (Jolly and Matthews 2017;

Lopez 2009; Pettigrew, Archer, and Harrigan 2016), suggesting that unschooling mothers are not demographically different to the wider body of blogging mothers. However, whilst most of the families might appear in many ways to conform to notions of normalcy in many aspects of their family life, the fact that they choose not to send their children to mainstream school at once places them as 'other' within a Western democratic society.

This alterity is not simply the negative act, or non-activity, of choosing not to attend school, but rather the myriad ways in which life is different due to non-conformist time and space. In other words, the non-institutionalisation of the children in a family alters the way in which the household – or specifically the non-attending children and their mother, relate to time and space. Theoretically, they do not need to orientate themselves to clock time (Nockolds 2017) in the same way as school-attending families, as they are not required to be in a specific location at a set time each day. Similarly, unschooling families are not legally required to be in a specific place, or take a specific journey each day. In other words, unschooling families do not become tied to a routine in the same way that families using school do (Lois 2010, 2017).

> When the children first came out of school, we tried to replicate their school day and did school at home. We tried to be structured – complete with a timetable and lesson plans. I was desperate to prove that I could do this. Desperate to prove that I could be a successful home-educating mama, and that I wasn't letting down my children in any way. I wanted to make sure that we fully covered everything that would have been covered at school.
>
> I failed.

Unschooled children are often also perceived as socially 'other'. Unlike children who attend school, unschooled children do not need to spend a significant amount of their term-time week in the company of unrelated children and adults – although this chapter will go on to show that often families make choices which mean that they will. The institutionalised engagement of children with large groups of non-relatives is typically thought of in contemporary Western democracies as socialisation, or an important part thereof (Medlin 2013). The perceived lack of socialisation of unschooled children is recognised by the blogging mothers, some of whom associate this perception as the key feature in marking their choice as strange. For example, one mother wrote that she was nervous about telling her friends she'd decided to home educate her child, and that the other mothers' perception of home education as strange was a driving factor in the deterioration and eventual end of those friendships.

Contrary to, in spite of, and perhaps as a result of, wider society's concerns over lack of socialisation of home-educated children, there are a broad range of social, educational and 'extra-curricular' activities such as sporting and leisure groups available to home-educating families (Murphy 2014). In

Manchester, UK, in addition to the home education groups run by home-educating families, there are specific regular sessions organised through forest schools, the Manchester museums and a city centre skate park, to name but a few.

> We've recently begun going along to a home education group run each fortnight locally. The kids love it. It seems they've really settled in, it's brilliant. We might join some other groups too as I'd really like to grow our social network a bit, and there is so much going on for home educators here in Gloucestershire.

Unschooling families appear to have a variety of motivations for using these groups, including creating some structures in their week, giving the children access to new activities or specialist equipment or training, and fostering and developing friendships. Growing your own social network is of course a clear counter to feelings of otherness, which might engender isolation. This idea will be returned to in the final section of this chapter. What is interesting to note in the mothers' discussions and perceptions of otherness is that in fact the act of not engaging with the national curriculum, or indeed, not engaging with any curriculum at all, is not typically discussed as a key axis of difference. Rather the difference concerns their use of time and space, and specifically their choice not to leave their children in the care of state school (Lois 2013; Kraftl 2016) – thus the alterity is recognised in their relationship to a state institution rather than the process of education itself.

Mother-child relationship

The blogs that I explored for this research were authored by women as the primary care-giver and educational facilitator. Thus, for the children and young people who are the focus of this research, the parent-child relationship that was particularly different for unschooled children compared to their mainstream-school counterparts was the mother-child relationship. This reflects the more generalised research concerning gender and home schooling that shows that home education is overwhelmingly a mother's responsibility (Lois 2017, 2009). Whilst in most of the families the mothers were either single parents or the fathers were at work during the daytime, there were two families in which the children did appear to have extended time with their fathers as a result of homeschooling. In one example a father worked shifts and was able to spend some time with the children during what would have been the school day. In another example the father took responsibility for the children one day a fortnight to give the mother a 'day off'. However, by and large, the decision to unschool appeared to primarily shape the mother-child relationship.

> Sometimes I feel like I'm attempting to do 'three full time jobs' at the same time, and failing at all of them. It's hard to write this as a home

schooler, because we're just supposed to 'suck it up' and get on with it. Because, really, it's our fault, isn't it. It's our choice. If we want to, we can put them in school, and end this.

In the unschooling families that were the focus of this research, the mothers took the key roles as both educators and carers. Whilst unschooling as an approach technically does not require much 'preparation' in terms of programmes of education and resources, mothers took the lion's share of the responsibility for strewing[3] and for facilitating activities at home and in the community. At first, the combined role of carer and educator may seem at odds with the normative mother-child relationship, primarily of care, assumed to dominate in neoliberal society. However, as discussed already, research shows that even where children attend mainstream school, and whether in single, female-headed households, or heterosexual two-parent families, it is typically the mother who takes primary responsibility for children's education (Barker 2011; Holloway and Pimlott-Wilson 2013). So whilst home education magnifies the role of the mother as educator within the family (Lois 2017), arguably this is an extreme on a scale, rather than a complete radical realignment of the mother-child educational relationship experienced in most mainstream educating households.

Conversely, the role of mother as carer within an unschooling family may at first appear to be simply an extension of the typical caring relationship found in most mainstream schooling households, given that research shows that mothers continue to be the primary caregiver to, and educational supporter of, their children (O'Brien 2007, 2005). However, it could be argued that in fact the geography of the caring relationship between mother and child in unschooling families is vastly different to that within mainstream schooling families, as mothers' time and space is experienced differently due to a lack of institutionalisation – and thus a non-reliance on clock time or required mobility – and enforced comobility. That is to say that for home-educating school-aged children, mothers are in the company of children, supervising and supporting their mobility, far more intensively than a mother would be for mainstream-educated children, who would be out of a mothers' immediate responsibility and control for at least six hours each working day. Therefore, home educating may have a very significant impact on a mother, not only in the lack of clock time and required mobility that are collateral damage in an institutionalised system of education, but in terms of intraembodiment. In other words, the being together and moving together of mothers and their home-educated children is likely to lead to a very different sense of maternal identity for a mother who is always with their children, in contrast to a mother that grants a school *loco parentis* for a portion of the day.

Children's agency

Another clear alterity, common to many of the blogs, was the way in which unschooling families are ordered and managed. The unschooling families featured in the blogs demonstrated largely less hierarchical family structures than might be assumed for a mainstream-educated family. That is to say that families appear to be, to a greater or lesser extent, democratic units in which children have a significant amount of agency – or at least greater agency than their mainstream-educated counterparts. The agency of children is a necessary part of unschooling which is grounded in an ethos of child-led decision making about what a child wishes to learn, and how they wish to learn it. Unsurprisingly, this child-led learning ethos does appear to spill over into other areas of family life, with children contributing to decisions about leisure activities and holiday destinations – or in the case of the two nomadic families, where to next. Linked to the unschooling ethos is the idea of gentle parenting, which similarly encourages parents to respect their child's autonomy and to encourage a child to make good decisions, rather than instructing them or forcing them to behave in a particular manner. On one hand unschooling may lead to more democratic parenting styles, however, I think it's more likely – given the enormity of the decision to unschool in the first place – that families that choose to unschool are typically practicing a more gentle form of parenting and prioritising children's agency and autonomy from the outset. This supposition is supported by the narrative in some of the blogs that begins well before the children are of compulsory school age, and as mothers begin to look for approaches to education that reflect their broader parenting ethos.

> I think no matter what we are doing it's important to make sure everyone is happy with the decision, for instance some days I might feel like staying in but it's apparent that the kids need to get out so I then change the plans and think of something that can maybe get us out of the house for the morning, meaning that they are then happy to be at home in the afternoon.

This level of agency in their temporality, or the rhythms or their day, is obviously in stark contrast to the experiences of mainstream-educated children who must spend their days in spaces and engaged with activities determined, regulated and assessed by professional adults (Barker et al. 2010; Catling 2005; Gagen 2015). Indeed, Kraftl (2016) finds that the conditioning of young children's minds and bodies through formal schooling, in contrast to the opportunity for autonomy and child-led practice at home, was a key driver in mothers choosing to withdraw their children from mainstream schooling. In contrast to many of the examples of 'otherness' in this collection, unschooling children, arguably marginalised in some ways, are in fact distinctly more powerful and experience more rights and freedoms, due to having more control over their day-to-day lives, than their contemporaries attending mainstream school.

Unschooling as identity and mitigating otherness

The preceding discussion has sought to demonstrate the ways in which unschooled children and their families might stand out as 'other' from families who engage with mainstream schooling. The discussion showed that it is not simply a value placed on the mode of delivery of education or on what the children are learning, but more broadly, the perceived difference hangs on a choice about engagement with one of the most widely accessed institutions in the world. An intentional alterity might be read into, or might be reflected by, this familial de-institutionalisation, spurring concerns from institutionalised families over the broader societal impact of failure to recognise the superiority of the state and teaching professionals as 'educator' that unschooling may reflect.

> At times we find ourselves the 'talk of the town' when out and about. Occasionally we are stopped and questioned about the children not being in school – and reactions to the home-educated response are mixed. Some people are curious, asking lots of questions and being quite positive and interested. Others are very negative and tell us how they find it hard to believe that we are allowed to do such a thing. I've also been told that I am ruining the futures of my children. I've learnt to ignore those that don't understand. It's not personal.

Choosing not to attend school means that unschooling families operate on a different time, and in different spaces to families who use mainstream schooling. However, of graver concern to both home-schooling families and outsiders is the ability to provide social opportunities, and to engage with the community, when the children are not attending school. A key means of mitigating this, discussed earlier in the chapter, is the use of home-education-specific social groups and special interest groups run by home-educating families and by organisations in the local community. Blogging mothers discuss these groups as both something to do, some routine in the week, and a chance to network and meet other home-educating families. These groups are seen as a social opportunity both for the children who attend and for their mothers, who form friendships and relationships of support with other parents also doing parenthood differently.

> I also discovered, learning more about the home education world, how many groups there are... other days we meet up with family and with friends. We really don't spend very much of the week at home!

In so doing, creating a network of other families who also educated at home, parents were able to create a sense of normalcy within a particular identity as a member of a subculture (Collom and Mitchell 2005). Given that there are many approaches to home education, and that unschooling is a particularly controversial sort of home education, it is likely that there are also tensions within these broad home education networks. However, interestingly, the mothers who blogged did not reflect on bad experiences within these groups. It may have been that they were aware that the blogs were a public space, and did not want to air their dirty laundry in public. Similarly, they may be keen to present home education as a friendly and happy community, to those non home educators reading the blog.

The mothers whose blogs I engaged with were also seeking to develop networks through online relationships. The manner in which this was done varied between blogs. Some mothers were using their blogs more clearly as diaries, recording events and experiences for their own sake, or because they felt they might be informative or interesting to others. Other mothers were more explicitly using their blogs as a way of providing information on home education to others. This included the practicalities of home education, such as deregistration, and giving information and providing hyperlinks to sites addressing the philosophy of unschooling. Through these blogs mothers were able to connect with other unschooling families, generating networks and online community – developing a sense of normalcy.

> Any questions? Please do feel free to ask away.
> Have you got any tips for activities to do outside with kids?
> Here's where you might go for more information about home schooling, and information you can trust.

This is likely to be particularly important for unschooling families who might find that they are considered more 'other' than other sorts of home-educating families who use a formal curriculum. The difference here lies in the fact that home educators following a curriculum retain an explicit recognition of the value of expert knowledge and therefore do not present such a direct challenge to the notion of the institution. By contrast, unschooling challenges the notion of required knowledge and unseats traditional approaches to adult-child relationships as a unidirectional flow of power from adult to child. At least superficially, this suggests a far greater threat to the education system, society and the economy more broadly, as this approach questions the right and the relevance of an adult-designed education system.

Conclusion

Unschooling children might be considered 'other' both in terms of the difference of their lived experience to those of their mainstream educated peers, and in terms of the way in which unschooled young people may be at risk of marginalisation due to perceived difference. However, unschooled young people are amongst the minority in this volume of being intentionally 'other' (see also Hall and Pottinger, this volume). That is to say that the alterity of their lifestyle is a reflection of personal or parental choice, rather than a legally protected aspect of their identity such as race, religion or impairment, or a reflection of familial disadvantage such as poverty, political instability, bereavement or a relationship breakdown. This chapter has demonstrated that, whilst unschooling may lead to families being viewed as 'other', unschooling may also be the product of political and philosophical decisions to raise children in a way that doesn't sync with children attending mainstream school. The parents in this study chose to address this alterity in two ways. Firstly, by looking outside their immediate neighbourhoods to form local, and online, social networks of other home-educating families. In these situations, unschooling families socialised alongside other sorts of home-educating families, finding comfort in support from others who chose to educate outside mainstream schooling options. The mothers who blogged also chose a second strategy to address alterity, presenting publicly accessible information to engage and inform a wider audience, intending to make unschooling appear less 'other' to the mainstream-educating outsider.

Notes

1 Registration is required for families who withdraw their children from mainstream schooling. However, whilst registration is typically encouraged by Local Education Authorities, it is not mandatory. This is due to a quirk in English law that sees home education as the default practice – i.e. families have to apply and register to send their children to school, and families who do not apply are presumed to be choosing to educate their children at home.
2 An interconnected network or list of blogs publicised through a well-regarded, often organisational, blog.
3 Whilst unschooling is child-led, it is common for families practising unschooling to make specific adult-led choices about the materials (such as books) available to children at home to ignite their interest in a variety of topics.

References

Barker, John. 2011. "'Manic Mums' and 'Distant Dads'? Gendered Geographies of Care and the Journey to School." *Health & Place* 17(2): 413–421.
Barker, John, Pam Alldred, Mike Watts, and Hilary Dodman. 2010. "Pupils or Prisoners? Institutional Geographies and Internal Exclusion in UK Secondary Schools." *Area* 42(3): 378–386. doi:10.1111/j.1475-4762.2009.00932.x
Campbell, Pamela Lou Rogers. 2012. *A Qualitative Analysis of Parental Decision-Making in Regards to Homeschooling*. East Lansing: Michigan State University, K-12 Educational Administration.
Catling, Simon. 2005. "Children's Personal Geographies and the English Primary School Geography Curriculum." *Children's Geographies* 3(3): 325–344. doi:10.1080/14733280500353019
Collom, Ed. 2005. "The Ins and Outs of Homeschooling: The Determinants of Parental Motivations and Student Achievement." *Education and Urban Society* 37(3): 307–335.
Collom, Ed, and Douglas E Mitchell. 2005. "Home Schooling as a Social Movement: Identifying the Determinants of Homeschoolers' Perceptions." *Sociological Spectrum* 25(3): 273–305.
Cooper, Bruce S, and John Sureau. 2007. "The Politics of Homeschooling: New Developments, New Challenges." *Educational Policy* 21(1): 110–131.
de Wit, Emma Emily, Daniel Eagles, and Barbara Regeer. 2017. "'Unschooling' in the Context of Growing Mental Health Concerns among Indian Students: The Journey

of 3 Middle-Class Unschooling Families" *The Journal of Unschooling and Alternative Learning* 11(22): 33.

Gagen, Elizabeth A. 2015. "Governing Emotions: Citizenship, Neuroscience and the Education of Youth." *Transactions of the Institute of British Geographers* 40(1): 140–152. doi:10.1111/tran.12048

Gaither, Milton. 2009. "Homeschooling in the USA: Past, Present and Future." *School Field* 7(3): 331–346. doi:10.1177/1477878509343741

Gaither, Milton 2017. "The Homeschooling Movement and the Return of Domestic Education, 1998–2016." In Gaither, M. (ed.), *Homeschool: An American History*, 241–305. New York: Palgrave Macmillan.

Griffith, Mary. 1998. *The Unschooling Handbook: How to Use the Whole World as Your Child's Classroom*. 2nd edition. Rocklin, CA: Three Rivers Press.

Holloway, Sarah L, Phil Hubbard, Heike Jöns, and Helena Pimlott-Wilson. 2010. "Geographies of Education and the Significance of Children, Youth and Families." *Progress in Human Geography* 34(5): 583–600. doi:10.1177/0309132510362601

Holloway, Sarah L, and Helena Pimlott-Wilson. 2013. "Parental Involvement in Children's Learning: Mothers' Fourth Shift, Social Class, and the Growth of State Intervention in Family Life." *The Canadian Geographer/Le Géographe Canadien* 57 (3): 327–336.

Holt, John. 1995. *How Children Learn*. Revised edition. Reading, Mass: Da Capo Press.

Holt, Louise, Sophie Bowlby, and Jennifer Lea. 2017. "'Everyone Knows Me .... I Sort of like Move about': The Friendships and Encounters of Young People with Special Educational Needs in Different School Settings." *Environment and Planning A: Economy and Space* 49(6): 1361–1378. doi:10.1177/0308518X17696317

Jolly, Jennifer L., and Michael S. Matthews. 2017. "Why We Blog: Homeschooling Mothers of Gifted Children." *Roeper Review* 39(2): 112–120. doi:10.1080/02783193.2017.1289579

Jolly, Jennifer L, Michael S Matthews, and Jonathan Nester. 2013. "Homeschooling the Gifted: A Parent's Perspective." *Gifted Child Quarterly* 57(2): 121–134.

Kraftl, Peter. 2013a. *Geographies of Alternative Education*. Bristol: Policy Press.

Kraftl, Peter. 2013b. "Towards Geographies of 'Alternative'Education: A Case Study of UK Home Schooling Families." *Transactions of the Institute of British Geographers* 38(3): 436–450.

Kraftl, Peter. 2014. *Informal Education, Childhood and Youth: Geographies, Histories, Practices*. New York: Springer.

Kraftl, Peter. 2015. "Alter-Childhoods: Biopolitics and Childhoods in Alternative Education Spaces." *Annals of the Association of American Geographers* 105(1): 219–237.

Kraftl, Peter. 2016. "Moments of Withdrawal: Homeschooling Mothers' Experiences of Taking Their Children Out of Mainstream Education." In Cameron, A., Dickinson, J. and Smith, N. (eds.), *Body/State*, 157–172. London: Routledge.

Kunzman, Robert. 2017. "Homeschooling and Religious Fundamentalism." *International Electronic Journal of Elementary Education* 3(1): 17–28.

Levy, Tal. 2009. "Homeschooling and Racism." *Journal of Black Studies* 39(6): 905–923.

Liberto, Giuliana. 2016. "Child-Led and Interest-Inspired Learning, Home Education, Learning Differences and the Impact of Regulation." Reviewing editor: Rebecca Maree English. *Cogent Education* 3(1). doi:10.1080/2331186X.2016.1194734

Lois, Jennifer. 2006. "Role Strain, Emotion Management, and Burnout: Homeschooling Mothers' Adjustment to the Teacher Role." *Symbolic Interaction* 29(4): 507–529.

Lois, Jennifer. 2009. "Emotionally Layered Accounts: Homeschoolers' Justifications for Maternal Deviance." *Deviant Behavior* 30(2): 201–234. doi:10.1080/01639620802069783

Lois, Jennifer. 2010. "The Temporal Emotion Work of Motherhood: Homeschoolers' Strategies for Managing Time Shortage." *Gender & Society* 24(4): 421–446. doi:10.1177/0891243210377762

Lois, Jennifer. 2013. *Home Is Where the School Is: The Logic of Homeschooling and the Emotional Labor of Mothering.* New York: NYU Press.

Lois, Jennifer. 2017. "Homeschooling Motherhood." In Gaither, M. (ed.), *The Wiley Handbook of Home Education*, 186–206. Hoboken, NJ: Wiley-Blackwell.

Lopez, Lori Kido. 2009. "The Radical Act of 'Mommy Blogging': Redefining Motherhood through the Blogosphere." *New Media & Society* 11(5): 729–747. doi:10.1177/1461444809105349

Medlin, Richard G. 2013. "Homeschooling and the Question of Socialization Revisited." *Peabody Journal of Education* 88(3): 284–297. doi:10.1080/0161956X.2013.796825

Morrison, Kristan. 2014. "Homeschooling as an Act of Conscientious Objection." *Journal of Thought* 48(3/4): 33.

Murphy, Joseph. 2014. "The Social and Educational Outcomes of Homeschooling." *Sociological Spectrum* 34(3): 244–272.

Nockolds, Danielle. 2017. "Working Sole Parents and Feminist Perspectives on the Intersection of Gender and Time." *Journal of Sociology* 53(1): 231–244. doi:10.1177/1440783316651496

O'Brien, Maeve. 2005. "Mothers as Educational Workers: Mothers' Emotional Work at Their Children's Transfer to Second-Level Education." *Irish Educational Studies* 24(2–3): 223–242. doi:10.1080/03323310500435513

O'Brien, Maeve. 2007. "Mothers' Emotional Care Work in Education and Its Moral Imperative." *Gender and Education* 19(2): 159–177. doi:10.1080/09540250601165938

Pettigrew, Simone, Catherine Archer, and Paul Harrigan. 2016. "A Thematic Analysis of Mothers' Motivations for Blogging." *Maternal and Child Health Journal* 20(5): 1025–1031. doi:10.1007/s10995-015-1887-7

Pini, Barbara, Kalervo N Gulson, Peter Kraftl, and Rae Dufty-Jones. 2017. "Critical Geographies of Education: An Introduction." *Geographical Research* 55(1): 13–17.

Ray, Brian. 2015. "African American Homeschool Parents' Motivations for Homeschooling and Their Black Children's Academic Achievement." *Journal of School Choice* 9(1): 71–96. doi:10.1080/15582159.2015.998966

Schafer, Mark J, and Shana S Khan. 2017. "Family Economy, Rural School Choice, and Flexischooling Children with Disabilities." *Rural Sociology* 82(3): 524–547. doi:10.1111/ruso.12132

Sherfinski, Melissa, and Melissa Chesanko. 2016. "Disturbing the Data: Looking into Gender and Family Size Matters with US Evangelical Homeschoolers." *Gender, Place & Culture* 23(1): 18–35. doi:10.1080/0966369X.2014.991703

Snee, Helene. 2012. "Youth Research in Web 2.0: A Case Study in Blog Analysis." In Heath, S. and Walker, C. (eds.), *Innovations in Youth Research*, 178–194. New York: Palgrave Macmillan.

Tollefson, Megan Nicole. 2007. "Social Isolation Meets Technological Innovation: Towards Developing a Model of Communication Among Parents Who Homeschool." PhD Thesis, Grand Forks: The University of North Dakota.

von Benzon, Nadia. 2019. "Informed Consent and Secondary Data: Reflections on the Use of Mothers' Blogs in Social Media Research." *Area* 51(1). 182–189.

Webster, Rebecca. 2013. "The Relationship Between Homeschooling and Child Abuse." Symposium on University Research and Creative Expression, Central Washington University, 14 May. https://digitalcommons.cwu.edu/source/2013/oralp resentations/137

Wilson, Helen F. 2014. "Multicultural Learning: Parent Encounters with Difference in a Birmingham Primary School." *Transactions of the Institute of British Geographers* 39(1): 102–114. doi:10.1111/tran.12015

9 Being seen, being heard

Engaging and valuing young people as political actors and activists

Sarah Marie Hall and Laura Pottinger

Chapter summary

This chapter draws on findings from a two year ethnographic and participatory project following *Team Future*, a young people's political campaign home-grown in Greater Manchester, UK. Focused particularly on agenda setting, engagement and listening, we offer insights into the value of, and potential challenges for, young people as political actors and activists. We also make the case for greater acknowledgement of age and generationality within debates about intersectionality, including how these meet with class and race in practices of and opportunities for political activism.

Introduction

Questions and concerns about visibility and audibility of young people in political activism and debates have grown in the contemporary age. Whether voicing discontent with gun control in the US, the Occupy movement in Hong Kong, or self-organising to address the UK's exclusionary, racist and ageist Brexit discourse, it would appear difficult for policy makers or the public to claim that young people are not interested in politics. Using the latter of these examples, in this chapter we draw on empirical findings and observations from a study tracing the everyday politics of a youth-led campaign called *Team Future*. The campaign began the day following the EU referendum, on 24 June 2016, when the majority result to leave the EU was announced, instigated by two young, black, working class men from Manchester. They were concerned about the lack of engagement with young people's life experiences, opinions and ideas for the future within both the Leave and Remain camps of Brexit, as well as within local and regional politics concerning devolution and austerity (also see *Team Future* et al. 2017).

The *Team Future* campaign quickly gained momentum, at its peak including over a hundred young people aged 14–18, and was provided institutional support by RECLAIM (www.reclaim.org.uk), a leading youth engagement charity based in Greater Manchester. Established in 2007, RECLAIM has developed a strong regional and growing national identity as a space for encouraging and developing working class young people's potential as political activists and campaigners. These ideas strike at the heart of contemporary concerns for human geographers, who have been paving the way in the social sciences on understandings of young people as more than political becomings, but as political beings in the 'here and now' (Mills 2017; *Team Future* et al. 2017). Capable of engaging, campaigning and debating, recent years have seen slow shifts in policy and public forums towards recognising the contributions of younger voices, although engaging with young people remains as much a photo opportunity as a genuine call for dialogue. Furthermore, much academic work to date on young people's politics involves researching young people from above, defining their political actions and the remits thereof, rather than working alongside or co-producing with young people (Percy-Smith 2010; Skelton 2010).

In what follows we explore how and why young people's views, experiences, ideas and priorities matter in political processes and decisions, using the example of our ethnographic and participatory research following *Team Future* from 2016–2018. After a brief overview of literature on young people's politics and intersectional issues, we identify how political issues and agendas are developed and led by young people and those who facilitate their engagement. Ultimately, we reveal how young people – particularly from working class backgrounds and black and minority ethnic communities – are being seen and being heard, notwithstanding the challenges they face to be taken seriously as political actors and activists. We indicate in our conclusions how academics, organisations and policy-makers might benefit from improving (and providing opportunities to hone and activate) their listening skills. As such, this chapter makes important contributions to understandings of alternative childhoods; not only in terms of the plurality and intersectionality of experience between young people, but also in understanding youth as a space of political possibility. As part of this, we implore scholars to more carefully consider the matter of age and generation in discussions of intersectionality, as an important element in the framing of structural and social inequalities.

Young people's politics: engagement, participation and the politics of intersectionality

Until recently, young people[1] have been notably absent from political geography (Bosco 2010). Following calls to address this exclusion and to '*em*power children and young people as… political actors in and beyond their daily worlds' (Philo and Smith 2003, p.112, their emphasis) a 'youthful political geography' has grown and matured over the last twenty years (Skelton 2013). A key achievement of this work has been to challenge dominant characterisations of young people as politically apathetic (Philo and Smith 2003; Gordon and Taft 2011). Though geographers and academics in related fields of youth studies and education have been successful in 'making visible'

(Skelton 2013, p.127; 2010) the political agency of young people in its diversity and complexity, the sub-discipline maintains a marginal position within the broader field of political geography. In the discussion that follows, we highlight significant currents within this literature.

Popular and academic narratives overwhelmingly represent the relationship between young people and politics in terms of a problem of declining interest and disaffection (Sloam 2011; Schubert 2017), often with reference to lower rates of participation in formal political institutions (Topping and Barr 2017). Consequently, broad debates over young people's political engagement are frequently channelled into narrow discussions about voting (a challenge we have encountered in communicating our own research, e.g. Hall and Pottinger 2017). The preoccupation with electoral politics is problematic in a number of ways. As we have argued alongside young activists and fellow academics, it denies the multifaceted ways in which young people *do* engage in society and are impacted by political processes operating at multiple scales (*Team Future* et al. 2017; Skelton 2013; Ruddick 2003; see also Willman, this volume). Significantly, those under 18 are excluded from participating in local and general elections in the UK at present. Conceiving political engagement in such narrow terms means that the voting practices of *young adults* (aged 18–25) come to provide evidence for the lack of engagement of *young people in general*, though the latter is a heterogeneous and ill-defined group including teenagers with varied legal rights and positionings (Evans 2008).

In a move that reinvigorated ostensibly radical debates around lowering voting ages in wider UK politics (Eichhorn 2014), the Scottish independence referendum in 2014 offered 16 and 17 year olds the opportunity to vote on Scotland's constitutional future. While extending suffrage might begin to address procedural intergenerational injustices (Lecce 2009) for *some* young people excluded from formal democratic processes, enfranchisement of 16 and 17 year olds does not challenge the demarcation of politics as an adult activity that only becomes accessible once a legally determined threshold of maturity is crossed. Furthermore, debates about lowering the voting age still rest on much less radical assumptions of young people as *a priori* apolitical and in need of training or encouragement. The notion of young people as presently disaffected is implicit, for example, in suggestions that stimulating interest in voting at a younger age could leave a 'participative footprint' of increased engagement into adult life (Mackie and Crowther 2015, p.5).

A focus on voting also propagates limited, adult-sanctioned delimitations of 'acceptable' political agency: typically acts that reinforce stability and the status quo (Staeheli et al. 2013; Gordon and Taft 2011). Responses to the 2017 UK General Election illustrate how young people's political agency is downplayed by sections of the political establishment even when it does take the form of electoral participation. Theresa May's decision to call a snap election saw a surge in youth voter registration and turnout, including 57% of those aged 18–19 (Curtis 2017), and particularly by supporters of left-wing Labour leader, Jeremy Corbyn. Commentators were quick to dismiss the emerging political engagement of younger demographics (e.g. Foges 2017), with Lord Sugar – entrepreneur and chief decision maker in the UK's 'The Apprentice' and 'Young Apprentice' – suggesting 'those people who voted for [Jeremy Corbyn] are quite bright and educated, but also not very experienced in life. I'm not sure if they really knew what they were voting for' (Boult 2017). Vested interests are served by perpetuating narratives of ill-informed, disengaged youth. As Millington (2016) notes with reference to the 2011 London riots, delegitimisation of young people's agency also extends to more critical strands of academic thinking that has dismissed moments of nascent youth politicisation as reformist, neoliberal or illegitimate (Žižek 2012; Harvey 2012; Rodgers and Young 2017). And yet, framing youth as politically disinterested detracts attention from the spatial, relational, and intergenerational exclusions that work to alienate young people in a variety of ways (Mannion 2007), perpetuating their marginalisation from debates and processes which affect their lives deeply (Bessant 2004).

A related yet distinct concern from that of young people's *politics* is youth *participation* (Skelton 2013). Ideas about the value of youth participation, understood as young people's active involvement in decision making at a variety of scales, continue to exert significant influence in youth policy in Western societies (Aldridge 2017; Bessant 2004; Clark and Percy-Smith 2006). Seemingly at odds with the stereotype of apathetic youth discussed above, youth participation is celebrated in mainstream policy as a mechanism for empowering under-represented demographics and improving the efficiency of a spectrum of services (Farthing 2012; Bessant 2003). Youth participation initiatives commonly include youth councils or forums operating within schools, as part of local government or in connection with a particular service or organisation.

Critics, however, suggest the 'orthodoxy' of youth participation is 'part of a reformist discourse' that 'overlooks the problem of young people's negligible political status' and fails to challenge the lack of opportunities to participate equally in other aspects of society (Bessant 2003, p.87; 2004). Strong critiques have identified the tokenistic and exclusionary limitations of participation initiatives for almost two decades (Cooke and Kothari 2002; Matthews et al. 1999; Bosco 2010; Van Wijnendaele 2014). As typically adult-organised channels for young people to express their opinions while receiving training in formal processes of decision making, youth participation initiatives are also 'adultist' in that they assume 'adults are better than young people and are entitled to act upon young people in many ways' (Checkoway 1996, p.13, cited in Gordon and Taft 2011, p.1511).

Geographers point out that youth participation is often conflated with consultation or 'having a say', reflecting an 'overwhelming focus' on voice in contemporary research and policy (Mills 2017, p.1; Wilkinson 2015; Clark 2017). As Percy-Smith (2010, p.110) notes, '"participation initiatives" tend to focus on young people expressing a view rather than being more fully involved in all phases of the planning and decision making cycle' that could potentially be extended to include 'reflective inquiry, decision making, action and evaluation'. Young people's voices are vulnerable to adult manipulation and can be recruited to script predetermined agendas within tokenistic forms of participation that prioritise passive 'listening' over learning and acting (Bessant 2004; Clark and Percy-Smith 2006).

Furthermore, as Kallio and Häkli (2011) note, official modes of participation such as youth councils are neither accessible to all (i.e. they often require a type of membership or official position, as well as travel resources for attendance), nor do they cover the full gamut of issues that are important to young people. Arguing that participation might be more productively operationalised if understood as 'a spatial practice involving the socio-spatial interplay of people and settings', Percy-Smith (2010, p.109) also draws attention to how 'the construction of spaces for participation (socially, culturally and aesthetically as well as physically) directly influences whether and how people participate', and Mannion (2007) foregrounds the relational and intergenerational factors shaping opportunities for inclusion.

While interrogating the intergenerational power dynamics in formal spaces of youth politics and participation is a pressing concern, there remains a tendency to consider adult and young people's politics as taking place separately, and this is reflected in the types of empirical setting in which research into young people's politics is conducted. Attention to the spaces in which adult and young people's politics overlap and intersect (Mannion 2007) might yet offer further opportunities to spotlight the exclusions and inequalities that maintain marginal positionings of young people, as well as identifying moments of intergenerational alliance and opportunities to address power imbalances.

It is here that questions around intersectionality can be usefully applied to address the particularities of young people's political engagement. While intersectionality has become something of a buzzword within contemporary geographical discussions, Hopkins (2017) notes that social geographers have been applying intersectional thinking since the 1990s (also see Hopkins 2018; O'Neill Gutierrez and Hopkins 2014; Vaiou 2018). By this, he is referring to the ways in which 'specific forms of inequality are mutually constituted' (Hopkins 2017, p.2). The activist and academic roots of intersectionality are considered important for when this idea is applied, so as to not erase the foundational work of black feminists and critical race theorists – such as Kimberlé Crenshaw, Leslie McCall, Angela Davis and bell hooks – who brought these ideas into public discussion (Hall 2018; Rodó-de-Zárate and Baylina 2018). These early writings were borne from the notion that 'race and gender couldn't be analysed as being neither mutually exclusive nor separate from each other, contributing to a more complex and dynamic understanding of social relations and power structures' (Rodó-de-Zárate and Baylina 2018, p.548). That is to say, it explores the ways in which social identities meet and – as the term would suggest – intersect, rather than being a 'single categorical axis' (Hopkins 2017, p.1) or a layering or adding up of inequalities (also see Brown 2012; Vaiou 2018).

The application of an intersectional lens was founded on the relationship between race, class and gender, as a particular constellation of identities; indeed, Crenshaw's early work focused on the experience of black working-class women in the US and their experiences of everyday oppression, differentiated because of their intersecting identities. However, it is important to note that intersectionality is not limited to a politics of identity; rather, it is rooted in a politics of activism against structural, symbolic and systematic oppression. There are thought to be three key types of intersectionality:

> Structural intersectionality is about the ways in which black women have to deal with 'multi-layered and routinized forms of domination' (Crenshaw, 1991: 1245) such as those associated with housing inequalities or employment practices. Political intersectionality focuses on the ways in which black women belong to at least two marginalized groups and so often have to engage with different political agendas. Representational intersectionality focuses on how images of women of colour – and debates about these – tend to overlook the intersectional interests of such women. (Hopkins 2017, p.2)

In this chapter, we touch on all three examples, though with a particular focus on the second – political intersectionality – although our focus shifts slightly from the traditional triad of gender, race and class, whereby age, race and class emerged as the foremost intersecting issues.

In light of this, it is worth noting that there have been discussions of late about the combinations of identity or 'array of vectors' (Hopkins and Noble 2009, p.815) that are explored within intersectional research. Brown (2012), for instance, argues that geographers have been 'extremely *uneven* in which intersections we have investigated' (p.542, emphasis in original), and in particular makes the case that 'beyond gender and race... other axes of identity and structures of oppression have received far less attention' (p.544). Hopkins (2018, p.587) similarly asks: 'is it appropriate to only consider the intersection of gender and race, or gender and class without necessarily paying much attention to the issues of age, disability, religion or sexuality?' These calls to broaden the analysis of 'different constellations of intersections' (O'Neill Gutierrez and Hopkins 2014, p.386) have led to the development of age, life-course and generation as key categories in the axes of power (see Hopkins 2017; Hopkins and Pain 2007). Notwithstanding, much of this work focuses on 'the intersections of gender and youth with other markers of social and cultural difference' (O'Neill Gutierrez and Hopkins 2014, p.384), with concerns that attention to race – arguably a key consistent of intersectionality as an approach – falls out of the bottom.

Furthermore, there is still a sense that age and inter-generationality are missing from intersectional perspectives, and that 'much work is still to be done with regard to the intersection of particular age, generational and other identities and their spatial specificity' (Hopkins and Pain 2007, p.290). While both authors have been involved with RECLAIM for a number of years (particularly Laura, having worked with RECLAIM from their beginnings), we were acutely aware of our positionality as employees at a university and being white women in our early thirties, doing research on the topic of youth politics. We therefore sought to ensure that our approach offered possibilities for intersectional engagement, with a varied and flexible methodological design. And so, in what follows, we draw upon ethnographic observations, participatory tasks with *Team Future*, interviews with RECLAIM staff and our own reflections from following the campaign for two years, to explore young people's political engagements, with particular attention paid to intersectional inequalities in agenda-setting.

Whose agenda? Exploring young people's political engagements

It makes sense to begin our discussion with when Sam and Elijah entered the RECLAIM offices on that morning in June 2016. Their concerns did not come from nowhere; both young men and their peers had been involved with RECLAIM for a number of years and in their various programmes that work to develop young people as political leaders. While RECLAIM's ethos is to work *with and for* rather than *on* young people, to develop person-specific goals and progress, much of their work before this had been led by organisational interests, often aligned with grant opportunities or local interest campaigns. *Team Future*, therefore, marked a break from these practices (common in academic research, too), from setting political agendas for young people, into something genuinely led by young people – and from inception. A member of staff from RECLAIM described the significance of these events:

> Sam and Elijah came in and spoke about racism, really, and intolerance in society. And I think there'd been that video on the news of the guy on the tram in Manchester being really racist to a guy and telling him to go home even though he was from Canada or something like that, and all that kind of ridiculousness. And they'd come in and they were, from what I hear, very angry but also very positive. And that's when the session started really, like what are we going to do? And in my opinion from then – I even thought before [the EU referendum] – we need to have a bit of a plan in place because either way, whether we stay or whether we go, there's something to be done here. [...] So we went from there with sessions, and they were quite complex at first I suppose, there were lots of things on the table. But I felt it kept coming back to: political leaders don't have the lived experience, they don't represent us, there's no diversity. And therefore there are issues with representation. It was very much about the types of political leaders they wanted. (Interview 2)

This desire for a youth-led movement was, therefore, couched as much in hopefulness and a desire for change as distressing personal experiences around everyday racism. Moreover, these experiences were not thought to be shared across society, and certainly not by figures who make political decisions on behalf of young people. Young people's politics are then inherently intersectional, because the idea of a lack of 'lived experience' in political leaders points not only to differences in social identity but also the structural inequalities that undergird society and political decisions-making (Crenshaw 1991; Hopkins 2017): a lack of diversity, voice and representation.

Having said this, it is here that the intrinsic complexities of young people's political engagement comes to the fore. Because while *Team Future* was envisaged and kick-started by two young men, it was because of their affiliation with an organisation such as RECLAIM and the intergenerational alliances (Mannion 2007; Gordon and Taft 2011) already developed between young people and RECLAIM staff that their efforts were able to flourish. One might argue that they would not have looked to political campaigning had they not already been involved – and this may be so – although our argument here is around the resources and capacities for young people to actually be engaged in politics. Key workers at RECLAIM (some of whom had been involved as young people and later became employees) were acutely aware of these tensions:

> The fact that two young people who are really like, politically engaged, and I think aware, partly because of what they've experienced with RECLAIM and having that space to engage and so on, then utilising that to then come to the organisation and then be like, 'you guys, would you do something about it, we need to do something about this.' […] knowing that you have someone that you feel comfortable with. Like, you can just go to and speak, speak to… […] what can happen sometimes at other places perhaps is, they do all the right stuff until it comes down to taking action, and that wasn't the case. Everything materialised and it was led by them and supported by RECLAIM staff. (Interview 3)

For *Team Future* to get off the ground, then, adult support and expertise were needed. For while the 'space to engage' offered a point of contact for young people to mobilise and develop their movement, this was still a workplace, the offices of an organisation. Formalised support that would not be available to young people was needed for the campaign – funding, materials, skills, advice, time, energy, and so on. This is not to undermine *Team Future* as a movement for change, but to identify what makes young people's politics unique. It also highlights how adult and young people's politics can overlap and intersect in complex ways (Mannion 2007).

These tensions around giving young people responsibility for campaigning whilst not overburdening them was inherent in our own research project, too, and reflects common discussions around the ethics and politics of research with this group (see Farthing 2012; O'Neill Gutierrez and Hopkins 2014). Part of our time was taken with observing *Team Future*'s activity, alongside a participatory element for supporting and co-developing the campaign's aims. Working with various combinations of young people, we agreed on a number of goals: to create a set of demands and a toolkit of best practice for those working with young people; to develop a short film featuring young people from *Team Future*; to organise a series of youth-led film screenings and workshops; and to co-write an academic article with young people (see *Team Future* et al. 2017). We of course had ideas from our own funding bid, with targets to meet and a funder to keep sweet, and a timescale to work within, but worked as flexibly as possible.

However, sometimes when young people made suggestions about the activities we were supporting, we had to impose boundaries or limits based on the availability of time, money or resources. The film could not, for example, be twenty minutes long because we did not have the funding to shoot extensive footage. Nor could we visit six cities in screening the film, or book large venues for the workshops, again due to funding and time restrictions. In tailoring the activities we were supporting, because they could not run independently of us or the resources we were offering, one might suggest that we shaped *Team Future*'s activities towards our own (academic) agenda.

Concerns about imposing agendas on young people was something that RECLAIM staff spoke very candidly about. For the young people in RECLAIM the limits placed around their political engagement were often as a result of intersecting obstacles regarding age, as well as class, race and ethnicity. These included practical constraints, such as the requirement for many young people to be in school or college during the working week. And while lived experience regarding intersectional inequalities is, in our view, a very powerful political tool, young people might also lack some of the knowledge and more diverse experience associated with adulthood. This is where challenges of genuinely youth-led activism and political campaigning emerged in our discussions with staff about agenda-setting:

> So for me, one of the biggest things facing working class young people is social housing and the prospect of there not being enough housing. But they might not have realised that yet, so it's kind of a difficult line to tread of what you think is a class issue for young people and what young people are saying. So, you don't want to say 'this is a working class problem, what do you think about it?'. You want to encourage that conversation anyway but there's not many ways of doing it without being the person to be like 'what about this?' It's the kind of fine line between being entirely youth-led where all of the ideas have come from them, or the way that people might look to an organisation like RECLAIM and be like 'what do you think about gentrification? What do you think about the lack of social housing in Manchester?' So it's a fine line. (Interview 4)
>
> I think it's made a challenge for us of, how do we manage quite an intense work stream of trying to get young people to go out and

campaign and enable that space to be truly youth-led, that we're not just like 'this is the campaign, go and be the face of it.' Actually getting them to come up with the campaign and the tactics and who they want to talk to... that's really hard. (Interview 5)

The second extract here highlights another important element in young people's campaigning, about young activists being exposed in new ways by being the 'face' of a campaign or movement, as well as requiring them to have the energy and motivation to really get behind a campaign, which is more likely when it is genuinely youth-led. This is not particular to youth-led organising, and could be said for any campaign group with a large grassroots support base. However, the idea of a lack of knowledge 'yet' or not being au fait with broader political agendas and climates (as in the first extract) was attributed to youth activism but at the site of intersectional axes where age meets class (Hopkins and Pain 2007; O'Neill Gutierrez and Hopkins 2014).

As a result of these multiple sensitivities in youth-led activism, when the idea for *Team Future* was raised, it led staff to reflect on their role and their place in the lives of the young people they worked with. Aligning with literatures that describe young people as political beings and becomings (see Mills 2017; Vanderbeck 2008; Wood 2012) staff described to us how they sought to listen to and represent young people:

> I'm a facilitator because I'm there to support the group to come up with their own ideas and to develop those ideas, so it's kind of action learning as well. (Interview 1)
>
> I think the way we work with young people is actually quite different, the way that we are very responsive and reactive to what's going on for young people and that we're willing to trust in them and enable them to take the lead in a way that a traditional service provider wouldn't. (Interview 5)
>
> I think that's a challenge, of what is your role in facilitating young people, youth-led campaigns. Because you want to step back as much as you can but you also don't want to see it flop. (Interview 4)

Common terms emerged in these discussions with staff, seeing their responsibility as being to 'respond', 'react' and 'facilitate' young people's political agendas, rather than lead or impress them. However, for young people's campaigning to work, as mentioned, resources are needed that might not be entirely youth-led, leaving adults feeling responsible for the outcome of campaigning activity. There was a sense that they could only step back so far, because they would be held to account if the campaign was inaccurate, inappropriate or outdated, and at times this led to staff feeling conflicted about imposing their own ideas or agendas.

The imposition of political agendas is not only an issue particular to youth-led political organising, but also in when their campaigns were 'released' into the world and in discussion with political actors. Being seen and being heard was a constant challenge for *Team Future*, and one which we regularly noted from our observations of meetings and events either attended or led by the young activists. One poignant example came from *Team Future*'s work on Devo-Manc, the short-hand term for Greater Manchester's devolution of political and (some) economic powers, overseen by an elected mayor. In early 2017 they had a series of meetings set up with various mayoral candidates, and an opportunity to tell these political leaders what mattered to them. The meetings, however, were for the most part a one-way conversation. The candidates all typically told the young people what they were going to do for them, rather than ask if they even cared about the issues in their shiny campaign manifestos.

One topic that came up regularly as an example of young people's views being ignored, to the point that it became something of a joke within *Team Future*, was the issue of bus passes. Candidates all centred cheap local transport and an integrated network within their campaigns, and yet for the young people this issue was low down on their agenda. They expressed this in their initial meetings with candidates, but to little avail. When Andy Burnham was elected as Greater Manchester Mayor, a key support staff for RECLAIM reached out to his office and secured a meeting with him. The purpose was for *Team Future* to discuss their concerns and 'asks', in their environment and at a time that suited them. A few days later, we asked some of the other young people how the meeting went:

> They said 'It was ok. Yeah. He was alright. He kept going on about bus passes for 16–18 year olds though. We tried to tell him about our project, but those bus passes, that was all he wanted to talk about'. (Laura, field diary, 5 December 2016)

As it turned out, the Mayor went ahead with the decision to launch half price bus passes for 16–18 year olds, which certainly addresses a significant issue – the cost of travel and ease of mobility for young people in Greater Manchester. However, the extract above underscores how young people's frustration and exclusion is still perpetuated by a failure to actually listen to their concerns, on their terms, and to press on with a particular political agenda. As a member of staff confirmed to us: 'I'd never heard young people talk about transport in the way that he's really talking about it, saying that it's a big issue and we need to tackle it' (Interview 4).

It became clear across our research that a key challenge for both young people and the adults facilitating their political engagement arose in navigating existing – and often conflicting – political agendas. There was an overwhelming feeling amongst *Team Future*'s young activists that when they were being seen and heard, their heightened visibility was often exploited for reasons other than a genuine desire for dialogue and collaborative change. They described feeling like 'election props' and 'photo opportunities', particularly by politicians, and told us repeatedly of their frustration at rarely seeing action as an outcome of their discussions. They felt as though they were pawns, utilised by candidates or leaders to prove their diversity credentials in an institution predominantly represented by white, middle aged, middle class men. They thought they were seen as useful and malleable, but not valued as serious political actors (also see Bessant 2004; Clark and Percy-Smith 2006).

One example sticks out for us, when in January 2018 a group of young people were invited to meet with Mayor Burnham. The meeting was called to be held in the council offices, with a request for a selection of young people from *Team Future* and RECLAIM to attend. We were asked by *Team Future* to go along too, and were interested in what might come of the meeting. The young people attending were excited and nervous, and spent many hours with RECLAIM staff preparing their asks for the Mayor. They had brought a copy of the *Team Future* manifesto – a document written collaboratively by young people across the project which outlined four strategic priorities for the group – and had chosen to focus on education, hoping to discuss the possibilities for education reform in Greater Manchester and the development of dedicated political education at secondary schools.

They were still practicing their pitches in the lobby while we waited to be called in. Once in the room, we were joined by two members of council staff who worked directly with the Mayor, but they did not introduce themselves or explain why they were at the meeting. The Mayor was late, a good twenty minutes behind schedule, and when he did arrive the conversation felt rushed. The group made their pitches, but the conversation was taken to where it had always been intended to go; the young people were not there to discuss their ideas or asks, but instead to be told about a new Greater Manchester Youth Combined Authority (GMYCA) that was to be formed. It would develop with or without their support; they were simply being informed and told about the opportunity. To be involved, they would have to complete a formal application outlining what they would contribute to the group, but the Mayor envisaged that two places would be taken by RECLAIM young people. 'Could these be rotating amongst different people?', members of *Team Future* asked. 'Maybe, but we want consistent engagement' was the official reply. The rules of the game had already been set.

While this may have been a genuine case of trying to engage with young people – meeting and talking to them, rather than sending an impersonal email to RECLAIM staff – it failed on a number of counts. Not only was the meeting called for, the young people's presence demanded, it was also held in an unfamiliar and intimidating environment (Percy-Smith 2010). The purpose of the meeting was unclear, leading to the young people's hopes being raised that they might be able to activate political change with this prestigious one-to-one conversation with the newly elected Mayor. Furthermore, the agenda was already arranged and immovable, rather than being agreed upon by all parties. The young people here were not even being consulted about their views, they were being given instructions regarding the GMYCA, itself an adult-sanctioned channel for young people's political expression modelled on established adult modes of doing politics (Checkoway 1996; Gordon and Taft 2011). A member of staff from RECLAIM summed this up perfectly: 'The biggest issues I think from what I hear from them… feeling powerless and that, also even if they have a voice they can shout all they want but actually still no-one's listening and there's no change happening' (Interview 2).

Conclusion

In this chapter we have worked through a series of examples from our ethnographic and participatory research with a young people's political campaign and the staff who supported this work to explore issues around how young people's agendas are being seen and heard. We structured this discussion according to the setting and delivering of political agendas, as a space in which issues around voice, authenticity and power were seen to play out. Youth activism was shown to be unique in a number of ways, particularly regarding the support needed to facilitate young people's political participation, leading to issues around the imposition of agendas. While young people, often in alliance with supportive adults who facilitated their engagement (Mannion 2007; Gordon and Taft 2011), were shown to be capable of setting and leading the agenda, their efforts were often unheard at best, or ignored or manipulated at worst. We posit that attention to intersecting inequalities is key here, for what they can reveal about political engagement in practice. For in the examples given above, it is significant that we are describing a group of young, working class activists, a significant proportion of whom also identified as black or minority ethnic. It is at the axes of race, class and generation that *Team Future* were striving to be heard and seen, to be taken seriously as political actors and activists, working against political institutions that are filled with people who have not walked in their shoes.

Seeing and hearing the *Team Future* campaign develop, and witnessing the issues above regarding agenda-setting being played out over time, certainly shaped our approach to the research we were conducting. We wanted to ensure that we were not repeating these same mistakes in our own participatory research, and that young people's political opinions and experiences were central to the activities we supported. In light of this, we want to close this chapter with some examples from young people on how to better work with them. This forms part of the demands, as mentioned earlier, which we worked with *Team Future* to develop and disseminate. They set out *Team Future's* ideas for including young people in decisions, debates and processes that affect their lives. Consisting of four elements – Collaborate, Value, Invest, Empower – we see the first of these themes, 'Collaboration' as particularly relevant to the discussions herein. *Team Future* demand that:

- Instead of talking at us / for us, work with us
- We're asking to be friends and to form relationships which will benefit us both

- Be straightforward with what you say and how you say it. We value clear communication, honesty and being treated as equal.

We call for other researchers hoping to support young people's political engagement to take heed of this, both in the political processes you research and in the way you approach working with young people. Please ask yourselves whether the agenda of your research could be shaped to better account for young people's political agendas.

Note

1 See Evans (2008) for a detailed discussion of how childhood and youth have been defined.

References

Aldridge, J. (2017) 'Introduction to the issue: Promoting children's participation in research, policy and practice', *Social Inclusion* 5(3), pp. 89–92.

Bessant, J. (2003) 'Youth participation: A new mode of government', *Policy Studies* 24 (2/3), pp. 87–100.

Bessant, J. (2004) 'Mixed messages: Youth participation and democratic practice', *Australian Journal of Political Science* 39(2), pp. 887–904.

Bosco, J.F. (2010) 'Play, work or activism? Broadening the connections between political and children's geographies', *Children's Geographies* 8(4), pp.381–390.

Boult, A. (2017) 'Lord Sugar: Corbyn voters "not experienced in life" and "didn't know what they voted for"', *The Telegraph*. www.telegraph.co.uk/news/2017/06/09/lord-sugar-corbyn-voters-not-experienced-life-didnt-know-voted/

Brown, M. (2012) 'Gender and sexuality I: Intersectional anxieties', *Progress in Human Geography* 36(4), pp.541–550.

Checkoway, B. (1996) *Adults as Allies*. Battle Creek, MI: W. K. Kellog.

Clark, A. and Percy-Smith, B. (2006) 'Beyond consultation: Participatory practices in everyday spaces', *Children, Youth and Environments* 16(2), pp. 1–9.

Clark, S. (2017) 'Voice or voice-over? Harnessing the relationship between a child's right to be heard and legal agency through Norwegian bullying cases', *Social Inclusion* 5(3), pp.131–147.

Cooke, B. and Kothari, U. (eds) (2002) *Participation: The New Tyranny?* London: Zed Books.

Crenshaw, K. (1991) 'Mapping the margins: Intersectionality, identity politics, and violence against women of color', *Stanford Law Review* 43, pp.1241–1299.

Curtis, C. (2017) 'How Britain voted in the 2017 general election', YouGov. Available at: https://yougov.co.uk/news/2017/06/13/how-britain-voted-2017-general-election/

Eichhorn, J. (2014) 'Newly enfranchised voters: Political attitudes of under 18 year olds in the context of the referendum on Scotland's constitutional future', *Scottish Affairs* 23(3), pp.342–353

Evans, B. (2008) 'Geographies of youth/young people', *Geography Compass* 2(5), pp.1659–1680.

Farthing, R. (2012) 'Why youth participation? Some justifications and critiques of youth participation using New Labour's youth policies as a case study', *Youth & Policy* 109, pp.71–97.

Foges, C. (2017) 'Let's stop treating the young as political sages', *The Times*. www.thetimes.co.uk/article/let-s-stop-treating-the-young-as-political-sages-fqv5b3cqg

Gordon, H.R., and Taft, J.K. (2011) 'Rethinking youth political socialization: Teenage activists talk back', *Youth & Society* 43(4), pp.1499–1527.

Hall, S.M. (2018) 'The personal is political: Feminist geographies of/in austerity', *Geoforum*. doi:10.1016/j.geoforum.2018.04.010

Hall, S.M. and Pottinger, L. (2017) 'Could young people show a new way for politics?' Policy@Manchester, http://blog.policy.manchester.ac.uk/posts/2017/05/could-young-people-show-a-new-way-for-politics/

Harvey, D. (2012) *Rebel Cities: From the Right to the City to the Urban Revolution*. London: Verso.

Hopkins, P. (2017) 'Social geography I: Intersectionality', *Progress in Human Geography*. doi:10.1177/0309132517743677

Hopkins, P. (2018) 'Feminist geographies and intersectionality', *Gender, Place & Culture* 25(4), pp.585–590.

Hopkins, P. and Noble, G. (2009) 'Masculinities in place: Situated identities, relations and intersectionality', *Social and Cultural Geography* 10(8), pp.811–819.

Hopkins, P. and Pain, R. (2007) 'Geographies of age: Thinking relationally', *Area*, 39(3), pp.287–294.

Kallio, K.P. and Häkli, J., (2011) 'Tracing children's politics', *Political Geography* 30, pp.99–109.

Lecce, S. (2009) 'Should democracy grow up? Children and voting rights'. *Intergenerational Justice Review* 4(4). doi:10.24357/igjr.4.4.510

Mackie, A., and Crowther, J. (2015) 'Informal learning experiences of young people during the Scottish independence referendum'. *Concept* 6(1), p.6.

Mannion, G. (2007) 'Going spatial, going relational: Why "listening to children" and children's participation needs reframing', *Discourse* 28(3), pp. 405–420.

Matthews, H., Limb, M., and Taylor, M. (1999) 'Young people's participation and representation in society', *Geoforum* 30(2), pp.135–144.

Millington, G. (2016) '"I found the truth in Foot Locker": London 2011, urban culture, and the post-political city', *Antipode* 48(3), pp.705–723.

Mills, S. (2017) 'Voice: Sonic geographies of childhood', *Children's Geographies* 15(6), pp.1–14.

O'Neill Gutierrez, C. and Hopkins, P. (2014) 'Introduction: Young people, gender and intersectionality', *Gender, Place and Culture* 22(3), pp.383–389.

Percy-Smith, B. (2010) 'Councils, consultations and community: Rethinking the spaces for children and young people's participation', *Children's Geographies* 8(2), pp.107–122.

Philo, C. and Smith, F.M., (2003) 'Guest editorial: Political geographies of children and young people', *Space and Polity* 7(2), pp.99–115.

Rodgers, D., and Young, S. (2017) 'From a politics of conviction to a politics of interest? The changing ontologics of youth politics in India and Nicaragua', *Antipode* 49(1), pp.193–211.

Rodó-de-Zárate, M. and Baylina, M. (2018) 'Intersectionality in feminist geographies', *Gender, Place & Culture* 25(4) pp.547–553.

Ruddick, S. (2003) 'The politics of aging: Globalization and the restructuring of youth and childhood', *Antipode* 35(2), pp.334–362.

Schubert, J. (2017) 'Out of work and low on enthusiasm: young Germans are tuning out of politics', *The Conversation*. https://theconversation.com/out-of-work-and-low-o n-enthusiasm-young-germans-are-tuning-out-of-politics-84392?utm_campaign=Echo box&utm_medium=Social&utm_source=Twitter#link_time=1506010927

Skelton, T. (2010) 'Taking young people as political actors seriously: Opening the borders of political geography', *Area* 42(2), pp.145–151.

Skelton, T. (2013) 'Children, young people, politics and space: A decade of youthful political geography scholarship, 2003–2013', *Space and Polity* 17(1), pp.123–136.

Sloam, J. (2011) 'Introduction: Youth, citizenship and politics', *Parliamentary Affairs* 65(1), 4–12.

Staeheli, L.A., Attoh, K., and Mitchell, D. (2013) 'Contested engagements: Youth and the politics of citizenship', *Space and Polity* 17(1), pp.88–105.

Team Future, Pottinger, L. and Hall, S.M. (2017) '"Have you heard that young people are RECLAIMING their future?" Towards a bold, ethical and hopeful politics of Brexit and beyond', *Local Economy* 32(3), pp.257–263.

Topping, A., and Barr, C. (2017) 'The received wisdom is that young people don't vote. Could that change?' *The Guardian*. www.theguardian.com/politics/2017/jun/07/ received-wisdom-young-people-dont-vote-could-that-change-general-election

Vanderbeck, R.M. (2008) 'Reaching critical mass? Theory, politics, and the culture of debate in children's geographies', *Area* 40(3), pp.393–400.

Van Wijnendaele, B. (2014) 'The politics of emotion in participatory processes of empowerment and change', *Antipode* 46(1), pp.266–282.

Vaiou, D. (2018) 'Intersectionality: Old and new endeavours?' *Gender, Place & Culture* 25(4), pp.578–584.

Wilkinson, C. (2015) 'Young people, community radio and urban life', *Geography Compass* 9(3), pp.127–139.

Wood, B.E. (2012) 'Crafted within liminal spaces: Young people's everyday politics', *Political Geography* 31, pp.337–346.

Žižek, S. (2012) *The Year of Dreaming Dangerously*. London, Verso Books.

Part III
Out of place

10 Young survivors of sexual abuse as 'children out of place'

Korinna McRobert

Chapter summary

This chapter looks at children 'out of place' as a category, focusing on the example of the sexually abused and thus sexually exposed child as experiencing an 'out of place' childhood. The issue is contextualised by looking at the origins of the term 'children out of place' and how the sexually abused and exploited child can be framed and categorised in this way. More specifically, understanding the 'place' of children in contemporary society is examined, mainly looking at the construction of childhood and its conceptual link to innocence. In order to explore childhood agency and innocence, the nature of the / a child-adult sexual relationship dynamic is analysed, in terms of one case represented in the film *UNA* (Doumanian et al. & Andrews, 2017), by Australian director Benedict Andrews, based on the play *Blackbird* (2005), written by British playwright David Harrower. The specific story, set in the UK and based on a true sequence of events of a relationship between an adolescent and an adult, is looked at in terms of the agency of the child, their ability or inability to consent, and their perceived innocence, as well as children's rights in general, using the framework of the United Nations Convention on the Rights of the Child (CRC; UN, 1989) to do so.

Introduction

Who are children 'out of place'? This is a term that was made known through the work of scholar Judith Ennew (see Conolly & Ennew, 1996). Since the International Year of the Child (1979) she became a children's rights researcher and activist, specialising in issues surrounding 'street children' and child sexual exploitation and abuse, reframing the way such children 'out of place' were seen, researched and considered in the fields of children's geographies, anthropology and childhood studies (see Beazley et al., 2014). She straddled the children's geographies and childhood studies disciplines and is represented centrally in both interdisciplinary fields. Originally the discourse surrounded 'street children', i.e. children who lived and worked on the streets (see Ennew, 1995). Over time, however, the term started being attributed to any child who does not conform to the expected, Western social identity of a child (see Invernizzi, 2017). This identity is generally one of a young person who is protected by adults, cared for by their parents, goes to school, does not work and is innocent, i.e. does not engage in sexual behaviour until the appropriate time, usually dictated by laws surrounding age of consent. Any child who engages in what society generally considers as 'adult behaviour' is considered 'out of place'. Examples of adult behaviour could be working, taking care of family members or being sexual (see O'Connell Davidson, 2018).

The example I wish to focus on in this paper is that of the *sexual child* as a child 'out of place'. By exposing children to the arena of sex through sexual abuse, I would claim that they are stripped of their assumed innocence, as they have been given knowledge of sexual relations. The issue of their own agency in the process will be looked at closely, highlighting the blurred boundary between childhood and adulthood (for a further discussion of this blurred boundary, see Markowska-Manista, this volume, in the context of child soldiers). Specifically, the focal point of this chapter will be on the child who has been sexually abused and exploited, in contexts involving *no* compensation – monetary or otherwise.

Codes of childhood

Ever since the philosophical, intellectual movement of the eighteenth century, which has come to be known as the Enlightenment, the European, Western world has been influenced by two main ideas surrounding childhood. The first dominant idea is that of the *tabula rasa*, i.e. of a child being a blank slate, empty and ready to be filled by knowing adults, coined by John Locke (see Locke, 1689). The second is the concept of childhood innocence that was popularised by Jean-Jacques Rousseau, who believed children to be born inherently innocent and only corrupted by outside forces (see Rousseau, 1762). These ideas framed the young person as passive and vulnerable (see Gunter, 2014). Children were and are seen as having nothing to contribute, since they have no inherent knowledge or ability, and are innocent to the ways of the world (see Montgomery, 2003). Thus, framing children as ignorant and innocent affords society's adults with the bulk of the (social and legal) power. The adult's responsibility thus became to educate, protect and cherish the young person's seemingly impressionable and fragile sense of self, thus, establishing disparate power relations between adults and children (see Liebel, 2004; Ennew, 1986), eventually leading to adults' disproportionate power over children's lives.

In other words, the association with children and childhood is predominantly a sense of innocence. When their innocence is debased in whatever form, children are seen as having 'lost their childhood' (see Montgomery, 2003). In other words, once a model of the ideal childhood is projected, its absence becomes not only obvious but a great loss. If something threatens any part of the conception of childhood, the child in question is marginalised, seen as having lost something integral. This public image of the child aided and validated changes in social policy (see Piper, 2000). For example, in Britain, following the ideology of the Enlightenment, there was a movement to ameliorate the conditions of child workers in factories, with the Ten-Hour Movement and the 1833 Factory Act (see UK Parliament, 2017). This ensured some regulation of the work done by children factory workers, established the minimum age of nine years and enforced two hours of schooling a day. Without the ideology of the vulnerable child needing more protection and care than an adult, there would have been no social move-ment to ensure that children had more regulations in their workplace. It also, however, established the image of a defenceless, weak and guiltless person (see Piper, 2000). This picture is one of non-agency, passivity and incompetence.

In order to further contextualise the sexual child as a 'child out of place', one can refer to scholar Sharon Stephens' (1995: 12) discourse on 'people out of place'. To begin to understand contemporary society's reaction to child-hood sexuality and childhood sexual abuse, one could draw the link between the social construction of childhood (see Ennew & Morrow, 2002) and expected innocence, as exposure to sexual knowledge through sexual abuse removes innocence, suggesting the loss of sentimentalised childhood (see Jacobson, 2001).

For the sake of this paper, childhood sexual abuse is defined as 'the sys-tematic abuse of power and need to control which is channelled through sex' (Sanderson, 2010: 17). Sex is not just defined as contact, penetrative or oral, but all the ways in which a child can be sexualised and used as a sexual object to gratify the adult. The damage done to the child is not only the sexual act itself but the lead up to it, the so-called grooming, the manipulation that accompanies it and the potential isolation that the child can experience through the stigma that this interaction can create. The childhood sexual abuse spectrum is very broad and does not only include coercion and force in order to involve a child in sexually exploitative situations (see Sanderson, 2010). The process can involve the building of a close, trusting relationship and then taking advantage of that trust to attain sexual access, often on a long-term basis. As a child / adult sexual relationship is illegal and socially taboo it needs to be secret, thus putting a lot of pressure on the child targeted. If the abuse is exposed, it could create *moral panic* (see Jacobson, 2001), leading to extreme legal and social interventions. Due to the difficulty in processing such traumatic events for the child and for those closest to the child, denial, hiding and disbelief may be used as coping mechanisms in order to deal with the situation. Short- and long-term effects of childhood sexual abuse can be severe and include post-traumatic stress disorder (PTSD), panic attacks, dissociation, memory loss, self-harming, eating disorders, fear of intimacy and general difficulty in building and maintaining relationships (see Sanderson, 2010).

As well as the trauma of the incident and the aftermath of it, abused children also have to deal with the burden of what childhood sexual abuse means to society. This indirectly stigmatises them, as well as the abusers, as they threaten society's definition of childhood. Such definitions are *institutionalised* through use, becoming socially and culturally constructed norms (see Oxford Dictionaries, 2017). Over time the stigmatisation and marginalisation of the survivors can add to the damage created by the abuse itself. The projection onto children by adults, in terms of what they should be, can thus eventually be damaging. However, this projection also serves as a safety blanket for the adults themselves, wishing to remember their *own childhoods* in a positive light, retaining the memory of their own childhood innocence (see Jacobson, 2001). Adults' nostalgia can thus burden children's present.

Codes of childhood sexuality

Following the Enlightenment movement, a code for childhood sexuality also developed in the West, with the focus on the innocence and malleability of young people (see Kehily & Montgomery, 2003). However, children did not necessarily conform to the sexless image that was projected onto them, as Victorian middle-class parents went to great lengths to stop their children from engaging in sexual activity. A diversity of tools and myths were implemented, including devices to stop children accessing their genitals, to stories about how masturbation can cause blindness (ibid.). In Victorian society, sexuality was seen as something that, if not limited and controlled, would threaten social order. One can go a step further and say that the state of childhood in itself *and* in combination with sexuality pose a level of threat. Sexuality and childhood can be said to occupy similar positions, in that they are assumed to be *natural* states but approached with caution (see Clark, 2013). By giving children no permission, social or legal, to be sexual, the power over their bodies lies with those who have permission to be sexual, i.e. adults. These individuals are thus allowed to regulate other individuals' behaviour, according to their own criteria. This creates an inherent imbalance in power (see Ennew, 1986), as well as misinterpretation of what it actually means to be developing sexually, what the needs and curiosities of children are, and how they can be heard and safeguarded. Sexualisation of children is generally seen as stemming from adult interference, not from the agency of the child themselves. This further separates the roles of the adult and the child in the child's sexualisation. Children are presented as incapable of being *responsible subjects* with their own agency in the context of sexuality (see Clark, 2013). The voice of the child in this discourse is absent. Even with the awareness of children's rights and the desire to represent children's lives and perspectives accurately, it is hard to do so if they are excluded.

In the 1920s, Freud suggested that children are born with an innate sexuality that develops over time but is not to be likened to adult sexuality (see Maywald, 2015). Freud's theory has been a big influence over the last 100 years in terms of how to view and frame childhood sexuality (see Piper, 2000). However, this still does not examine a child's active involvement in their sexual development and sexuality, nor does it take into consideration any external and social factors influencing the process of constructing the sexual identity of a child. The problem is that framing children as innocent, pure and angelic can give them a *special* attractiveness and sexuality that is secret and illicit, creating a taboo (see Kincaid, 1992). The innocence that is constructed to protect children ends up eroticising them, leading to *fetishisation*. Fetishised subjects are commonly associated with sexual charge / sexuality and infantilism, highlighting the attractiveness of something framed as helpless. The attraction is thus rooted in a contradiction between helplessness and power (see Schroeder, 2008). Adult attraction to children can be seen as a symptom of fetishisation of the pure, the clean and the unthreatening.

Sexuality and agency

The term *agency* is used to mean one's ability to make choices, and exercise self-will and determination (see Thrift, 2014). In terms of any scientific and academic perspective on childhood and children, one needs to consider the child's own agency in every context affecting them. Researchers in the fields of childhood studies (see Nieuwenhuys, 2013) and children's geographies (see Costello & Duncan, 2006) are encouraged and expected to consider issues of child participation and voice. Children's agency in the context of their sexuality, however, is highly restricted through laws dictating minimum ages of sexual activity and consent, as well as social expectations of a child as a non-sexual and innocent being. The world of the child and the world of the adult are not meant to have a bridge between them when it comes to sexuality and sex. Children are meant to have sexual relations with nobody and adults are meant to have sexual relations with each other. Sexual relations between children are taboo, let alone relationships between children and adults. The voice of the child regarding their own sexuality is generally not seen as valid (see Rendel, 2000a). Transgressions are treated as crimes under most international rules of law. In these legal cases, the adult would be the perpetrator and the child would be the victim. The perceived agency, and therefore the presumed responsibility, will lie with the adult. The child is seen as someone who is not capable of making an informed decision in this context. There is no space conceptually for self-willed, self-initiated sexual activity on the part of the child (see Piper, 2000). If a sexually abused child would be seen to have had some agency in the starting and / or maintenance of the relationship that compromised them, the social and legal system would not be able to function as it is. This said, the legal system in the UK would assess situations also in relation to the age of the child involved. There would be different sentencing for an adult who has sex with an 11-year-old and one who has sex with a 14-year-old. And even in cases where a child under the legal age of consent actively presented themselves as someone 'of age', the adult who has sex with them is still held accountable and the act is considered 'statutory rape', irrespective of the agency of the child in the situation.

It is preferred by the legal system to see the situation as a straightforward scenario, where adults know exactly what they want and how to get it, and children are passive, naïve and unknowing. The sometimes unique and complicated relationships are not analysed in order to understand them but in order to make strong accusations. The power structures are not looked at intricately, framing the child as a powerless victim, needing protection from adult eroticisation (see Voléry, 2016), instead of being seen as a contributor to the dynamic they are part of. Society ought to be obliged to recognise, respect and try to understand the child's position, without assuming that their position is one of complete disempowerment. Stripping responsibility inadvertently strips agency. Of course, in a society where children are generally physically smaller, less educated, dependent financially and lacking the same legal rights as adults, the child is automatically in a disempowered position (see Montgomery, 2010). This does not mean, however, that they do not have some *thin agency*, known as the ability to act within highly restrictive situations (Klocker, 2007).

In order to contextualise children's sexual agency in terms of UK law, specifically within the realms of the National Health Service (NHS), I wish to briefly draw attention to what is recognised as *Gillick competence*. Since the case of *Gillick v West Norfolk and Wisbech AHA [1986] AC 112*, a person under 16 years may have the ability to access and consent to medical treatment, specifically in relation to contraception, without the knowledge of their parents or guardians. This means that a sexually active child can exercise their own agency to avoid disease and pregnancy. Being provided with advice and treatment surrounding contraception is not granted by default, however. A child has to undergo an assessment by a healthcare professional in order to prove they are capable in this context, known as being *Gillick competent* (see Rendel, 2000b). Criteria, also known as the *Fraser guidelines*, include: the child exhibits a desired level of intelligence and maturity, thus understanding the nature and ramifications of the treatment; the child could not be convinced to tell their parent or guardian and is highly likely to engage in sexual activity with or without the contraceptive treatment; the child's mental or physical health will probably deteriorate without the treatment or advice; the treatment or advice is in the child's best interests (General Medical Council, 2008: 40). However, a general practitioner (GP) is also obliged to follow a protocol to exclude sexual abuse. The GP would need to establish whether the child is sexually active, under what circumstances and with whom (see GPnotebook, 2018). If there is a suspicion of child sexual exploitation the doctor is obliged to alert the authorities. Thus, the legal system, as it regulates the UK NHS, *does* attribute children with agency and even the ability to consent to sexual activity, provided they have proven to also have the capacity to make such decisions for themselves.

When it comes to instances of sexual abuse and exploitation, however, it is problematic to attribute the child with agency. In doing so, one can be seen as blaming the child for what happened to them. Since an adult having sex with a child is illegal and thus criminalised, the roles of criminal and victim need to be attributed to the involved parties. Within a legal context this may be appropriate but in a social context, where an abused child needs to go through confronting legal proceedings, therapy and rehabilitation (see Crown Prosecution Service (CPS), 2017), the label of 'victim' may not help them recover and have the normal, post-abuse life that they may wish for. The innocence that gives children a special status in society also limits and silences them, making them invisible in the procedures set up to meet their best interests (see Cook, 2009). Somehow innocence *and* agency cannot inhabit the same space. This points to the predominant problem in our adult-led society, of adults having the power to make decisions for children. When talking about infants, for example, parents or guardians will have to make all decisions regarding the welfare of the child in their care. However, this means that important decisions concerning children's lives can be made without consulting them or trying to understand their motivations and perceptions, precisely because they see them as non-agentic beings. The voice of the child is not given priority in a debate that needs their central participation (see Clark, 2013).

In cases where the child does not identify as being coerced, it may be more fruitful to find out where the child's motivation and drive to engage in this kind of relationship came from, instead of trying to find ways of denying them their agency in this context. This could point to a need that has not been met for the child, through their society, their situation and / or their personal relationships. Each case would of course need to be looked at individually, as guidelines for dealing with such cases may be helpful but cannot be generalised and applied to everyone. Sexual agency and permission to sexuality can be said to be something that one earns, as one gets older, creating a kind of *generational order* (see Qvortrup, 2009) in terms of access to and etiquette in the world of sex, specifying age groups and their particular roles and statuses (see Voléry, 2016). Like other aspects of life, sex is perceived culturally, and framed appropriately, depending on social codes. But sex is also accepted to be a part of nature. The hybridity of nature and culture causes a paradox and contradictions that may be present when trying to define and present any sexual situation. Sex remains an *intersection* of nature and culture (see Paglia, 1992), a biological act put into cultural and thus subjective terms when described.

Analysis of adult-child sexual relationship in film *UNA*

In order to explore central issues surrounding childhood sexual abuse in the context of an intimate, long-term relationship between an adult and a child, I wish to analyse the film *UNA* (2017), directed by Benedict Andrews, which depicts such a situation. One does not encounter this theme often in the context of mainstream cinema and when childhood sexual abuse is presented, it generally depicts the child as emotionally destroyed and mentally unwell, focusing more on the heinous act rather than the lead up to it, the relationship behind it, or the complicated feelings surrounding it. For example, in the French 2013 Claire Denis film *Bastards* (Chioua et al. & Denis, 2013), the child is portrayed as a lost cause, institutionalised after being raped by her father. She is a two-dimensional character who is just a symbol for the loss of innocence that our society expects from a child who has been exposed to adult sexuality. It is rare to find a film that portrays the child involved with their dignity intact, attributing them a sense of agency and awareness of their situation, giving them a voice to explore the complexities of their predicament.

The film *UNA* portrays an agential child, nodding to the notorious tale of *Lolita*, the story of the erotic relationship between a professor and his pubescent step-daughter, attributing the child with an active role in the situation (see Harris & Kubrick, 1962). *UNA* is based in and filmed in the UK. It is written by British playwright David Harrower, based on his play *Blackbird*, using real events as the foundation for his script. I wish to use this contemporary piece of mainstream media to talk about a taboo topic. It helps to anchor the discussion in a pre-existing structure and story that can be followed by the reader, as well as giving a reference that can be viewed in combination to this piece of writing. This discussion will particularly explore the depiction of the adult / child relationship as a disruption to the traditional perpetrator / victim dichotomy and the agency this might appear to lend to the child in the relationship. *UNA* revolves around a relationship, between Una (female child) and Ray (male adult). The story is told from Una's perspective, both when she was a child and later as an adult. The film alternates between the present day and the time of their relationship when Una was 13 and Ray is presumed between 30 and 40. He was her neighbour, and friend to her father. In the present day, approximately 15 years later, Una is a young adult and Ray is middle aged. The 'then and now' device used is effective in communicating a fuller picture of the events in a narrative way, offering the child and adult perspectives in alternate. The film begins by introducing Una as a 13-year-old at her house, cut with her as a young adult out at a nightclub, leading back to the same house, where she still lives with her mother. The following morning, she drives to a factory and walks in, looking for Ray. After showing his co-worker a photo of him from the newspaper, she is told that the name of that man is in fact Peter. She gets taken to him all the same. Ray excuses himself from his work and takes Una to a public kitchen-come-locker room, where they can talk. The plot is mainly based on the confrontation between Una and Ray in his workplace, before moving finally to his home. In this time, she challenges him, starting a long dialogue about their history, broken up by flashbacks, showing the establishment of their relationship, how it led them to run away together and Ray's eventual arrest, followed by his prosecution. We are told that Ray went to prison for four years. Una went back to her life, forever changed and stigmatised by the event. In the present day, Ray is a free man with a new identity, a new life and the ability to hide from his past. This is presented in strong contrast to Una's inability to hide from the ongoing ramifications of this childhood relationship.

The aftermath of Ray's arrest and subsequent sentencing is described in great detail by Una, who needs Ray to know how bad her life got since then: 'I hate the life I've had... I wanted you to know that' (Doumanian et al. & Andrews, 2017). The way she delivers her story is strikingly lucid and self-assured. The role of the helpless and damaged victim is not attributed to her, even though she is clear about believing Ray abused her. She is determined, straightforward and active in making sure Ray hears her. One assumes that in all the years that she was cut off from him, her life remained at a standstill and what she knew she needed was indeed this confrontation, initiated by her. There is a sense that she feels safe in her decisions and trusts herself. The movie is a high-tension chamber piece framing the lead up to the end of Una's disempowerment. Una is given agency throughout the piece, as an adult, and also as a child. We are told about the notes she leaves Ray on his car windshield, about the secret codes they used to communicate with each other and her willingness to run away with him. Her delight in knowing how much she could sexually pleasure him was also something she mentioned (see Doumanian et al. & Andrews, 2017). This suggests that her own will and power were exercised throughout the interaction, blurring the boundary between childhood and adulthood, in terms of sexuality.

Despite Una's agency, Ray's higher social status as an adult male and evident manipulation skills are factors to be considered. He describes her as 'headstrong', 'determined' and 'wise beyond her years', which seems like he is trying to remove her from the category of the *child* and place her in the category of the *adult*, thus exonerating himself of responsibility and of the identity of a paedophile. He says, 'you were so impatient, sick of being treated like a child' (Doumanian et al. & Andrews, 2017), which can be interpreted as an attempt at convincing her she was as responsible and active in the establishment and maintenance of the relationship as he was. Ray argues that he was not attracted to Una's assumed innocence but instead to her maturity, insisting repeatedly through the film that he wasn't attracted to children in general, but just to her. Una, however, sees through this attempt to frame the relationship in a way that downplays the taboo of being in a relationship with a minor, as she retorts that all thirteen-year-olds feel tired of being treated 'like children' (see Doumanian et al. & Andrews, 2017).

Society sacralises and maintains childhood as a safe space characterised by dependency and lack of responsibility (see Cook, 2009). This is excused due to children's physical, intellectual and sexual immaturity, thus giving them an inferior social status. To clearly look at the power balance in a relationship between people with differing social statuses is challenging. It seems necessary to point out the power imbalance but also important to still attribute the person with the lower social status with agency. The filmmakers bravely take on this challenge, clearly demonstrating Una's evident willing participation, and even manipulation of Ray, at the time of the abuse. However, this nuanced and complex account clearly indicates that Una's consent, whilst freely given, was not 'informed'. That is to say that decisions arrived upon by a thirteen-year-old were then considered by that same person, fifteen years later to be 'stupid' and thus not valid or trustworthy. Thus, Una herself suggests that whilst she had power in the relationship, she should not have done. Through Una's adult eyes it had been Ray's responsibility to recognise her immaturity and to reject her advances.

UNA in terms of children's rights

Within the narrative of the film, Una and Ray's case was dealt with following due process, within the context of UK law. Ray was arrested by the police and charged with sexually abusing a child. Una was forced to comply with the investigation. She was questioned and had to undergo a medical examination to ensure that she had indeed had sex with Ray. Despite her wishes not to cooperate, she was sedated and examined anyway. The tragic irony is that she seemed to give consent to having sex with Ray, but she did not give consent to have her body examined. The legal system, which was attempting to protect and ensure her right to be free of sexual exploitation, went against her wishes in respect to her own body. She was rendered agency-less through this process. This leads one to observe how adults possess the rights and obligations to dictate decisions that affect and change children's lives, whether in the role of the protector or the abuser. Children are also attributed rights, however one can question how effectively they are able to exert these rights, as a child is a mere 'subjectivity', whose voice is only defined, controlled and explained by adults (see Loizidou, 2000). If the child has rights, it is only within the space that the adult gives them to exercise these rights and entitlements.

When using the United Nations Convention on the Rights of the Child (CRC) as a reference for children's rights, one can look at the dialogue between a few relevant Articles. Article 34 urges that State Parties, through their own laws, must protect a child from all sexual exploitation and abuse by preventing their engagement in '*unlawful* sexual activity' (italics my own). It is implied that sexual relations are harmful and against the child's consent. It is not stated that legal action shall only be taken if the child involved *wants* to take legal action, because they feel it is appropriate to their case and their feelings surrounding it. It also implies that intervention must comply with the law and, up to a point, the values of each State, which vary and are subjective.

The CRC grants children participation rights in Article 12, the right to express their views freely and be heard on matters that affect them; Article 13, freedom of expression and access to information; and Article 15, freedom of association. However, according to Article 3, all actions concerning children should treat the best interests of the child as 'a *primary* consideration' (italics my own). So, rights granted in Articles 12, 13 and 15 are nullified, as Articles

3 and 34 are prioritised. Simply listening to a child's explanation and expression of views or feelings without taking them into serious consideration risks becoming tokenistic, suggesting that adults can only tolerate and work with expressions from children that do not challenge them or the status quo of society in general (see Nieuwenhuys, 2013). Without some dialogue with the child involved, the action taken could only *assume* their best interests. Children are not seen as equal partners in the assessment of their own situations and problems and thus not seen as part of finding solutions either. It can also be assumed that, at least in some cases, children are seen to be unaware of their best interests and cannot be trusted to make the right decisions for themselves.

In Una's case, her right to freedom of expression and right to be heard, or not heard as the case may be, were denied. The protocol of the trial was that Ray was in a courtroom and Una needed to be in her lawyer's office. She was obliged to make a statement to a camera in front of her. This live stream was shown in the courtroom, where people there could see and hear her, but she could not see them. She was not allowed to see or talk to Ray, let alone send him a message, as she requested. After being denied that, she says what she wanted to say to the camera anyway. She asks him why he left her and tells him she loves him. Una never gets a response or reaction to her statement. This highlights the fact that the court hearing is something she did not want and was as much enforced on her as it was on Ray. The chance to communicate feelings and viewpoints may be important to the individual but is complicated when it comes to legal proceedings. This policy, based on the UK laws of the time, to not allow such expression and communication, is adopted in order to protect and safeguard the child, but it may end up ostracising them.

Article 39 obliges State Parties to ensure proper post-abuse recuperation, recovery and care of the 'child victim'. Una talks about her life after the trial: she went back to the same house; her mother just wanted to 'hush it all up' and pretend it never happened; her father was deeply angered by it and died a few years later in an accident; Ray's girlfriend attacked her in the street while she was out with her mother (Doumanian et al. & Andrews, 2017). Una was clearly not given proper care. Ray was put in jail, served a sentence and was out of her life. The responsible adults probably interpreted this as the end to her suffering. She makes us realise that it was the beginning of it. Her entire treatment was based on adults assuming what was best for her.

An interpretation of this story is that Una was mature enough at the time of her relationship with Ray to consent and only perceives the relationship as abusive in retrospect, because of the reaction of outsiders to their relationship and the gravitas of its aftermath for Una. The trauma she experienced could have been more from the intervention to protect her than from the relationship she was being protected from. In the relationship she played an active role but in the dissolving of it she was overpowered by those who were given the responsibility of acting in her 'best interests'. Much like the legal protocol described here, the CRC is often criticised for being a treaty drafted by State representatives who all share the common belief that children are passive, dependent and overall powerless individuals receiving rights courtesy of adults, not their own contributions (see Ennew & Morrow, 2002). However, in confronting Ray, adult Una says that she was too young and immature at 13 to make the informed decision to get into a relationship with him (see Doumanian et al. & Andrews, 2017), leading us to consider that an adolescent's decision-making abilities may not have reached their full capacities and children cannot be seen as fully agentic in a sexual relationship with an adult.

Everything Una was denied the right to talk about, she expresses in her adult interaction with Ray. When her family, community and society fail her, putting her 'out of place', she takes matters into her own hands, dealing with it in a way that is best for her. As an adult she has the right to assess what her best interests are. Una's plan to exorcise herself of the burden of this history does not end with a mere talk with Ray. She is determined to expose Ray, or Peter as he is now known, as a paedophile. She asks him if she was the only 13-year-old girl he slept with. He says yes. She seduces him at his workplace and even though they have sex he is unable to climax. She asks him if she is too old for him now. Ray flees the scene and goes home to his wife and prepares for a party. There is a brief scene of him and his wife having sex in their bedroom, which Ray is able to conclude with his eyes closed.

Una finds a way to get to Ray's new house, crashing the party. She meets his wife and explores the house. She finds a girl's room and meets a 13-year-old girl. She is Ray's stepdaughter. This alludes to the archetypal sexual constellation of the stepfather and the stepdaughter, also featured in *Lolita* (see Harris & Kubrick, 1962), highlighting the child-adult sexual dynamic and reminding us of Ray's status as a paedophile. Una confronts Ray, publicly. He engages with her, defensive and apologetic, exposing himself to all his peers, co-workers and wife. Once this is done Una walks away from the scene. There is one more shot of her as a 13-year-old and the film ends. Una can now leave this all behind. Her job is done. She has finally achieved justice for herself. Justice was not achieved for her through the law but through her own initiative.

Conclusion

It is fair to say that the sexually abused and the sexually exposed, as well as the sexually active, child are categories that do not fit the ideal model of childhood. The ideal model of childhood is measured in terms of the intactness of a child's innocence, and also ignorance, on the subject of sex. Protection from this sort of perceived harm is expected by the family institution and society as a whole, as reflected in the laws of the State. Thus, children exposed to sexual practices are seen as 'out of place'.

This category serves to point to issues surrounding children's lives and their rights within their situations, as well as critique the sometimes limited definition of *childhood*. In order to avoid potential generalisation surrounding the different 'out of place' categories it is advisable to look at all individuals involved as having their own specific agendas and agency. After the assessment of a situation where a child has been sexually abused, it would be preferable to look at each individual case carefully and specifically, in order to limit abstractions, generalisations and expectations of what it actually means, for that individual, to be sexually abused. By focusing also on ways of integrating the abused child back into society, rather than only protecting them from the harm of sexual abuse, their exclusion and marginalisation could be avoided in the long run.

References

Beazley, Harriot, Bessell, Sharon & Waterson, Roxana (2014). Sustaining the energy: A celebration of the life of Judith Ennew. *Children's Geographies*, 12(1). pp. 118–125.

Chioua, Brahim, Clerc, Laurence, Thery-Lapiney, Olivier (Producers) & Denis, Claire (Director) (2013). *Bastards* [Motion Picture]. France & Germany: Wild Bunch & Real Fiction.

Clark, Jessica (2013). Passive, heterosexual and female: Constructing appropriate childhoods in the 'Sexualisation of Childhood' debate. *Sociological Research Online*, 18(2). doi:10.5153/sro.3079

Conolly, Mark & Ennew, Judith (1996). Introduction: Children out of place. *Childhood*, 3, pp. 131–145.

Cook, Daniel Thomas (2009). Editorial: When a child is not a child, and other conceptual hazards of childhood studies. *Childhood*, 16(1), pp. 5–10.

Costello, Lauren & Duncan, Duane (2006). The 'Evidence' of sex, the 'Truth' of gender: Shaping children's bodies. *Children's Geographies*, 4(2), pp. 157–172.

Crown Prosecution Service (CPS) (2017). *Child Sexual Abuse: Guidelines on Prosecuting Cases of Child Sexual Abuse*. London: CPS. Retrieved on 11 February 2019 from www.cps.gov.uk/legal-guidance/child-sexual-abuse-guidelines-prosecu ting-cases-child-sexual-abuse

Doumanian, Jean, Daly, Patrick, Amsellem, Maya (Producers) & Andrews, Benedict (Director) (2017). *UNA* [Motion Picture]. United Kingdom: Jean Doumanian Productions & West End Films.

Ennew, Judith (1986). *The Sexual Exploitation of Children*. New York: St. Martin's Press.

Ennew, Judith (1995). Outside childhood: Street children's rights. In: Franklin, B. (ed.). *The New Handbook of Children's Rights: Comparative Policy and Practice*. London: Routledge, pp. 201–215.

Ennew, Judith & Morrow, Virginia (2002). Releasing the energy: Celebrating the inspiration of Sharon Stephens. *Childhood*, 9(1), pp. 5–17.

General Medical Council (2008). Children and young people. In: GMC (ed.) *Consent: Patients and Doctors Making Decisions Together*. London: GMC, p. 40.

GPnotebook (2018). Guide to consulting with a sexually active child. In: *General Practice Notebook – a UK medical reference*. London: GPnotebook. Retrieved on 11 February 2019 from www.gpnotebook.co.uk/simplepage.cfm?ID=x2016060775520544321

Gunter, Barrie (2014). *Media and the Sexualisation of Childhood*. London & New York: Routledge.

Harris, James B. (Producer) & Kubrick, Stanley (Director) (1962). *Lolita* [Motion Picture]. United Kingdom & United States: Metro-Goldwyn-Mayer.

Invernizzi, Antonella (2017). Understanding children's sexual exploitation and protecting human rights. In: Invernizzi, A., Liebel, M., Milne, B. & Budde, R. (eds.) *'Children Out of Place' and Human Rights. Children's Well-Being: Indicators and Research.* Cham: Springer International Publishing AG Switzerland, pp. 137–156.

Jacobson, Maxine (2001). Child sexual abuse and the multidisciplinary team approach: Contradictions in practice. *Childhood*, 8(2), pp. 231–250.

Kehily, Mary Jane & Montgomery, Heather (2003). Innocence and Experience. In: Woodhead, M. & Montgomery, H. (eds.), *Understanding Childhood: An Interdisciplinary Approach.* Chichester: Wiley, pp. 221–265.

Kincaid, James R. (1992). *Child-Loving: The Erotic Child and Victorian Culture.* New York/London: Routledge.

Klocker, Natascha (2007). An example of 'thin agency': Child domestic workers in Tanzania. In: Panelli, R., Punch, S. & Robson, E. (eds.) *Global Perspectives on Rural Childhood and Youth: Youth and Rural Lives.* New York/London: Routledge Taylor & Francis Group, pp. 83–94.

Liebel, Manfred (2004). The economic exploitation of children: A theoretical essay towards a subject-oriented praxis. Unpublished paper in association with book: Liebel, M. (2004). *A Will of Their Own: Cross-Cultural Perspectives on Working Children.* London & New York: Zed Books.

Locke, John (1689). *An Essay Concerning Humane Understanding.* London: Thomas Basset.

Loizidou, Elena (2000). *Lolita* at the interface of obscenity: Children and the right to free expression. In: Heinze, E. (ed.) *Of Innocence and Autonomy: Children, Sex and Human Rights.* Aldershot/Burlington: Dartmouth Publishing Company Limited/ Ashgate Publishing Company, pp. 124–139.

Maywald, Jörg (2015). *Sexualpädagogik in der Kita: Kinder schützen, stärken, begleiten.* Freiburg/Basel/Wien: Herder.

Montgomery, Heather (2003). Childhood in time and place. In: Woodhead, M. and Montgomery, H. (eds.), *Understanding Childhood: An Interdisciplinary Approach.* Chichester: Wiley, pp. 45–84.

Montgomery, Heather (2010). What is a child? What is childhood? In: Liebel, M. & Lutz, R. (eds.) *Sozialarbeit des Südens, Band 3: Kindheiten und Kinderrechte.* Oldenburg: Paulo Freire Verlag, pp. 21–29.

Nieuwenhuys, Olga (2013). Theorizing childhood(s): Why we need postcolonial perspectives. *Childhood*, 20(1), pp. 3–8.

O'Connell Davidson, Julia (2018). Thinking of local, global and globalized childhoods through the lens of slavery. Personal communication: Seminar presentation, November 6. Norwegian Centre for Child Research at the Department of Education and Lifelong Learning, Norwegian University of Science and Technology.

Oxford Dictionaries (2017). *English Oxford Living Dictionaries.* Oxford: Oxford University Press.

Paglia, Camille (1992). *Sexual Personae: Art and Decadence from Nefertiti to Emily Dickinson.* London: Penguin Books (Original work published in 1990 by Yale Press).

Piper, Christine (2000). Historical constructions of childhood innocence: Removing sexuality. In: Heinze, E. (ed.) *Of Innocence and Autonomy: Children, Sex and Human Rights.* Aldershot/Burlington: Dartmouth Publishing Company Limited/ Ashgate Publishing Company, pp. 26–45.

Qvortrup, Jens (2009). Childhood as a structural form. In: Qvortrup, J., Corsaro, W.A. & Honig, M.-S. (eds.) *The Palgrave Handbook of Childhood Studies*. Basingstoke: Palgrave Macmillan, pp. 21–33.

Rendel, Margherita (2000a). Sexuality and the United Nations Convention on the Rights of the Child. In: Heinze, E. (ed.) *Of Innocence and Autonomy: Children, Sex and Human Rights*. Aldershot/Burlington: Dartmouth Publishing Company Limited/Ashgate Publishing Company, pp. 49–63.

Rendel, Margherita (2000b). The United Nations Convention on the Rights of the Child and British legislation on child abuse and sexuality. In: Heinze, E. (ed.) *Of Innocence and Autonomy: Children, Sex and Human Rights*. Aldershot/Burlington: Dartmouth Publishing Company Limited/Ashgate Publishing Company, pp. 64–76.

Rousseau, Jean-Jacques (1762). *Emile, or On Education/ Émile, ou De l'éducation*. Paris: La Haye, J. Néaulme.

Sanderson, Christiane (2010). *The Warrior Within: A One in Four Handbook to Aid Recovery from Childhood Sexual Abuse and Violence*. London: One in Four.

Schroeder, Jonathan E. (2008). Fetishization. In: Donsbach, W. (ed.) *International Encyclopedia of Communication, Volume 4*. Oxford: Wiley-Blackwell, pp.1803–1808. Retrieved on 10 March 2017 from www.academia.edu/3068327/Fetishization

Stephens, Sharon (1995). *Children and the Politics of Culture*. Princeton: Princeton University Press.

Thrift, Erin (2014). Agency. In: Teo, T. (ed.) *Encyclopedia of Critical Psychology*. New York: Springer Science and Business Media, pp. 62–68.

UK Parliament (2017). *Living Heritage: Reforming society in the 19th century. The 1833 Factory Act*. London: HM Stationary Office.

United Nations (1989). Convention on the Rights of the Child. In: UNICEF (2009), *The State World's Children Special Edition: Celebrating 20 Years of the CRC*. New York: UNICEF, pp.74–83.

Voléry, Ingrid (2016). Sexualisation and the transition from childhood to adulthood in France: From age-related child development control to the construction of civilizational divides. *Childhood*, 23(1), pp. 140–153.

11 Realising childhood in an Urdu-speaking Bihari community in Bangladesh

Jiniya Afroze

Chapter summary

This chapter presents a discussion of childhood in an Urdu-speaking Bihari camp in Dhaka, Bangladesh. Drawing on data from ethnographic fieldwork, this chapter sheds light on children's experiences of everyday violence to understand how childhood is constructed and understood across generations. The chapter interrogates the dominant discourses and narratives on the ideologies of 'good child' and 'bad child', and the normative lens through which childhood is understood within an intergenerational frame. Shedding light on the perspectives of both children and adults, this chapter illustrates the complex ways in which children construct, perform and negotiate their agency, and puts emphasis on the importance of having a flexible and situated understanding of children's lives to realise the nuances of children's experiences in the context of everyday violence.

Introduction

> Shuhash, a 14-year-old boy, is in a classroom with his friends. The teacher comes and asks for the homework. The teacher beats Shuhash as he has not prepared his lessons. The teacher even tells him that if he does not do his lessons tomorrow then she would ask him to stand outside the classroom, under the sun. After school, Shuhash cannot join his friends at the football ground as he must finish his lessons before he goes to work in the afternoon. He is late for work as he struggled to finish his lessons. At work, the *mohajon* [supervisor] scolds Shuhash for being late to work. At the end of the day, he cut his payments for being late. When Shuhash hands over the money to his mother, she yells at him for bringing less money home. At night, Shuhash has a very disturbed sleep. He wakes up a few times screaming, 'Madam, don't beat me, I will do my lesson', '*Ustad* [Sir – to address the supervisor], I will not be late at work again', 'Ma, forgive me, I will not do this again'. (Extracts from field notes taken on 2 October 2016).

Drawing on data from a doctoral research project with children and adults living in an Urdu-speaking Bihari camp in Bangladesh, this chapter sheds light on children's experiences of everyday violence to understand how childhood is constructed and realised across generations. At the outset of this chapter I present an extract from my field notes concerning a drama production that I observed in my research site, to which I return later in the chapter. The drama was performed by some of the children in the camp as part of the celebration of children's rights week organised by a non-governmental organisation (NGO). The storyline does not speak for one particular child at the camp community, rather, the instances presented in the drama echo with the everyday life experiences of several participants in my research site. There was not much surprise or shock in the audience's response to the drama, as the issues presented are part of children's 'routine, inescapable, and mundane' experiences of everyday lives (Wells and Montgomery 2014, p. 1). Indeed the violence children experience in their everyday lives are not 'graphic and transparent' (Scheper-Hughes and Bourgois 2004, p. 2), rather they are so embedded in the everyday practices that often they are not commonly recognised as violence.

The discussion in this chapter is framed by briefly providing an overview of the historical rootedness of violence in this research context, then combining this with children's narratives of their experiences which are embedded within the gendered and generational relationship of power. While doing this, it interrogates the dominant discourses and narratives on the ideologies of 'good child' and 'bad child', and the normative lens through which childhood is realised within an intergenerational frame. Underpinning the discourses of understanding childhood from an intergenerational perspective is the belief that relationships between children and adults are not 'fixed'. 'Child-adult relations' are explored as 'relational processes' which go through 'constant change' across diverse contexts (Mayall 2012, p. 350). Shedding light on the perspectives of both children and adults, this chapter illustrates the complex ways in which children construct, perform and negotiate their agency.

Methods

The ethnographic fieldwork for this research was conducted between April and November 2016 in an urban slum in Dhaka, which is popularly known as a camp, referred to as *Bithika* camp in my research.[1] *Bithika* camp, one of the 116 camps in Bangladesh, has an area of 55,000 square feet with an estimated population of 3000 people. Data was collected from children between the ages of 5 and 18, and their parents and caregivers, through participant observation, semi-structured individual interviews, group interviews, and group discussions using participatory tools including community walking, hand puppets, photos, and vignettes. My initial access to the research site was made possible through my professional connections with a national NGO, *Oikotan*, which runs a project within the camp supporting families to promote children's rights. For the initial few weeks of data collection, my base was primarily the NGO drop-in centre in the camp, which allowed me to observe the

NGO-run activities as a way of familiarising myself with the people and their culture, and gradually establish trust with them. Slowly, using snowball sampling, I immersed myself in the everyday lives of the children and their families across the camp, including joining them at their homes, workplaces and play spaces, as well as their communal meeting spaces, i.e. tea stalls, water collection points and lanes.

Conceptualising childhood and violence

At the core of the sociology of childhood is the fundamental belief that childhood is socially constructed, children are social actors, and that children's views are central in understanding childhood (James et al. 1998). With the growing realisation that the experiences of children's lives are diverse rather than universal (James et al. 1998, Montgomery 2009), and that the meaning of childhood and adulthood changes across generations (Alanen 2001, Mayall 2002, Punch 2007), scholars in children's geographies in particular have explored ways of understanding childhood across time and spaces (Matthews and Limb 1999, Kraftl et al. 2012, Holloway et al. 2018). As mentioned earlier, the belief that both childhood and adulthood are socially constructed categories situates both children and adults in the contexts of their everyday lives, where the intergenerational interaction and relationships are better realised by exploring relationships within the given time and social spaces (Hopkins and Pain 2007). Building on the concepts of spatiality as raised by James et al. (1998), scholars focused on children's experiences of 'everyday spaces in and through which children's identities and lives are made and remade' (Holloway and Valentine 2004, p. 11). Many scholars also put emphasis on the 'the wider processes, discourses and institutions' which connect to each other (Ansell 2009, p. 191), in order to have a contextual understanding of how children negotiate power in experiencing everyday violence (Matthews and Limb 1999, Holloway and Valentine 2004, Pells and Morrow 2018).

A budding body of research in children's geographies recognises that children's experiences of violence are often embedded in their everyday use of space and place (Winton 2005, Parkes 2007, Bartlett 2018). Everyday violence appears to be an emerging approach to understanding children's lives (Wells and Montgomery 2014, Parkes 2015, Pells and Morrow 2018). As opposed to physical violence, which is primarily directed from one individual to the other, Galtung conceptualised violence in association with structural inequalities and divergences in the power dynamics within social structures (Galtung 1969). Building on Galtung's work, Farmer (2004) later elaborated on the social processes and arrangements, engrained in historical, economic, and political practices, which produce and reproduce inequalities and harms, e.g. poverty, exclusion, and injustice, to certain social groups over the other. Bourdieu (2004), on the other hand, brings out the concept of symbolic violence, where the relationship of the dominants and the dominated are not physical or coercive, rather it is more of an invisible battle of power, which instigates those who are dominated to internalise and legitimise the inequalities by normalising their everyday experiences of domination. Scholars reinforce the notion that violence is embedded in 'ordinary everyday lives' (Parkes 2015, p. 4), and thus to have a nuanced understanding of violence, it is important to break the binary of structural and symbolic violence, and to dig deep into the everyday 'nexus of power, violence and the control of space and place' (Bartlett 2018, p. 6) by investigating how power governs within families and communities.

The camp: historical context

The residents of this camp – commonly known as *Urdu-bhashi* (Urdu-speakers) or Biharis – have their origins in the Indian state of Bihar, and a few other Muslim-majority states in northern India, who moved to then East Pakistan – present-day Bangladesh – during the partition of the Indian subcontinent in 1947. Mostly due to the linguistic connection of the Biharis with the predominantly Urdu-speaking political elites of West Pakistan, the Biharis were associated more with West Pakistan, and were considered by many Bangladeshis to be collaborators with West Pakistan during the liberation war of Bangladesh in 1971. The language movement in 1952 – a movement to have Bangla as the national language as opposed to Urdu – laid the foundation for Bangali nationalism, and sowed the seeds of the independence of Bangladesh, which later overshadowed the linguistic identity of the Urdu-speaking Biharis in independent Bangladesh. Hence, immediately before and after the independence of Bangladesh, the Biharis became targets for revenge by some Bangalis, which led many Biharis to leave their homes and take shelter in 'temporary camps', supported by the International Committee of the Red Cross (ICRC).

Following the independence of Bangladesh, many Urdu-speaking Biharis continued to live in those camps 'with deep-rooted uncertainty, poverty, trauma, self-pity and hopelessness' (Paulsen 2006, p. 55), whilst other Biharis were 'repatriated' to Pakistan, although it was a country they had never belonged to. The political tensions between Pakistan and Bangladesh in the newly independent Bangladesh delayed the repatriation process which led some Biharis to demand their rights to citizenship in Bangladesh, whilst others pursued repatriation. The yearning for repatriation among many Biharis reinforced the Bangalis' view of the Biharis as a 'collective political voice that is "pro-Pakistani"'(Redclift 2010, p. 314), which was an affront to their Bangali nationalism (Basu et al. 2018). Consequently, in independent Bangladesh, many Biharis experienced *othering* in two ways: firstly they were 'othered' by the Bangalis because of their Urdu linguistic profile and their perceived affiliation to (West) Pakistan. Secondly, ideological differences that developed among the Biharis broke the sense of 'community' and drove them to experience dichotomies of 'self and othering' amongst themselves. In

Bangladesh, the Biharis managed to find employment either in *Benarashi Saree* businesses,[2] hand embroidery work (e.g. *karchupi, jardozi, jari* and stones) or worked as barbers and butchers, for which they were particularly trained. However, for many years, they experienced discrimination in accessing basic services and formal employment for not having a right to citizenship in Bangladesh, accompanied by the social stigma due to their linguistic identity, which they continue to experience even today. In 2008, 37 years after the independence of Bangladesh, the Bihari people received the right to citizenship. Nonetheless, the deep and long-standing historical roots of violence in the form of discrimination still have a deep impact on the lives of the Bihari people in the camp, as many of them are still trapped in the cycle of poverty and violence, explored in the following sections.

Experiencing '*julum*': everyday violence within a socio-spatial context

As Scheper-Hughes and Bourgois (2004, p. 1) articulate, violence is not only physical, rather it consists of 'assaults on the personhood, dignity, sense of worth or value of the victim'. The 'routine misery' of children in this research is largely generated by the baggage of 'deprivation, frustration and humiliation' that their families and community are carrying with them for many years (Bartlett 2018, p. 5). The legacy of socio-political deprivation and their denial of citizenship status in Bangladesh until 2008 have positioned them as a vulnerable community in the socio-economic context of urban Bangladesh. The legacy of social and spatial deprivation that the Biharis are carrying with them also entraps them in a 'continuum of violence' (Scheper-Hughes and Bourgois 2004, p. 1) which triggers individuals and groups to produce and reproduce violence within their contexts.

Binoy, 42, a father of three children, expressed his frustration and anger as he told me that many Bangalis address the Urdu-speaking Biharis as '*maowra*'[3] (descendants of the Maurya Empire) and '*rajakar*'[4] (collaborators) which he finds pejorative. Binoy was utterly exasperated as, even four decades after the independence of Bangladesh from Pakistan, the Bihari community is experiencing '*julum*' (violence) by the Bangalis. The '*julum*' that Binoy refers to is not any physical violence or attack but the everyday marginalisation and discrimination that they experience. Several participants in this research told me that they experienced discrimination in accessing education, getting a job, and having the right to vote, which eventually framed a generation which they consider is creating 'damage' within and beyond the camp. The everyday *julum* inhibits children from becoming 'capital' to the society, forcing them to remain as a 'burden' instead. As Arun, 36, said: 'we are damaging the country… We are unintentionally damaging our children – we have no other way around'. Binoy perceived this as his failure as a father, as he said, 'we [the Biharis] are confused… we are even unable to raise our children properly… I haven't been able to do my duties as a father'.

Becoming 'rotten' vs. being 'good'

Within this culture of violence, which is rooted in the social, political and economic histories of the camp, conceptualising childhood, and examining how children experience it, becomes complex. Central to ideas about childhood within the camp are the concepts of '*bhalo*' [good] children, as opposed to '*noshto*' [rotten] or '*kharap*' [bad] children, which are embedded in the concept of 'rottenness' associated with the camp itself. This resonates with what Balagopalan (2014, p. 87) portrayed in her research with street children in Kolkata, where the children expressed their desire to become '*manush*' [human], which exhibits how they want to fit themselves into the 'hegemonic form of personhood' yet 'embracing life as it presented itself to them'. To isolate themselves from the rottenness of the camp, the participants in my research expressed their desire to become '*bhalo*', as they expressed their fear that the 'rotten' environment of the camp could turn children 'rotten' too. Many of the adults also spoke about the need to counteract and challenge the prevalent discourse of the camp being a 'rotten' place by altering the rotten image of themselves, and their children, and turning it into something decent and respectable.

One aspect of the rottenness of the camp was expressed by mothers in a group discussion in which they expressed their anxieties and fears about the common, if underground, business of drugs and alcohol which prevents their children using public spaces with confidence. They feared that their children would become rotten if they witnessed violence or badness in the form of drugs or alcohol in their everyday lives. Koli, 30, said, 'I always keep my eyes on him. If I see him hanging around with other children, I drag him home'. Koli explained why she does so, saying, 'because the *mahal* [neighbourhood] is not good, the "society" [community] is not good, and this time is not good'. In a group discussion a few fathers reflected that when children roam around the camp, mostly unsupervised, they often fall into bad company; children get involved in carrying or selling drugs without realising what they are doing. Harun said, 'there are young boys who randomly roam in the streets. Even you give them 50 taka and ask them to bring *ganja* they will give you a *puri* [small file of *ganja*]. Ultimately, there is no doubt, by the time a boy is 15–16 years old, he would fall into a *neshar jogot* [the world of drugs]'. Adults often take advantage of children's day-to-day poverty and vulnerability by exploiting them into drugs and engaging them in hazardous labour (Atkinson-Sheppard 2017), which eventually push children in the camp into the social construct of 'rotten' child.

Children living in the camp are often labelled as '*camp er chhele-meye*' [camp children] – a communal category – and this affects the way children are positioned, and also position themselves, within the dominant constructs of childhood in the camp. Often, to stand away from this banal categorisation, children claimed that they actively positioned themselves as 'good children' and tried to fit themselves to the criteria of 'good child' set by the community.

Chakraborty (2009, p. 442) illustrated how young women in a slum in Kolkata negotiated the 'socially constructed ideal' of 'good Muslim girl', and challenged the hegemonic discourses of 'good girl' in relation to their use of public spaces. In many interactions with the children in my research, I heard them being very critical about those who do not conform to the ideals of the good child and further risk the reputation of the camp.

> Mosharrof: The *poribesh* [environment] here is not great. All *aaul-faul* boys.
> Researcher: What is *aaul-faul*?
> Rahul: It means *kharap*.
> Mosharrof: [asks the researcher] don't you know the meaning? They always like *ure-ghure* [loiter around]. We also roam around but...
> Naveed: [interrupting Mosharrof] ...they go to different places and do *gaala-gaali* [swearing and bad words].
> Mosharrof: They do *gaala-gaali*, they do *maara-maari* [hitting] – it doesn't feel good.
> Shimanto: We don't do *gaala-gaali*. If they hit at some point, then I feel irritated, only then I hit. If they swear then there is no point in swearing back. It will provoke fights. Then I just hit them.
> (Extracts from group discussion, boys' group)

The extracts from the discussion among the boys illustrate that there is a sense of 'self' and 'otherness' among children as they were distancing themselves from those who do not conform to the ideals of 'good boys' that is collectively realised among the participants. There is a consensus in the group that the boys who swear, use bad language, and hit others are perceived as 'bad boys'. Even though the participants in the above extracts mentioned that they also hit other children, they indicated that they do this as a defensive mechanism, which they do not count as something 'bad'.

In an attempt to fit into the communal construct of 'good children', a few participants said that they prefer to distance themselves from those who fall into the category of 'bad children' in the camp. This was also reflected in the work of Dunkley and Panelli (2007) as they illustrated young people socially distance themselves from their peers on the basis of their perceptions and understandings of what is good and bad. The following account of 14-year-old Adi reflects how he socially distances himself from other children in the camp while socialising with boys other than his ethnic group outside of the camp.

> I no longer get along with other children from the camp. Now all my friends are from a higher class [outside the camp]. I started to embrace their lifestyles... I like to spend money. So that others have the impression that I come from a decent family. I always want my friends to look at me with admiration... But my parents are really poor. I don't tell my friends

that I live in this camp, you know, people in this camp and even my parents are uncultured. (Adi 14, male, individual interview)

Some of the parents also expressed their preferences to remain '*nitol*' [clean] and '*niribili*' [isolated] from 'others' within the camp as a way to get away from the stigma of 'rottenness' which is associated with this place. Sabita, for example, made up her mind that she would arrange her children's marriages within Bangali families instead of in Bihari families, which would allow her children to get rid of the stigma they experience due to their ethnic identities. Sabita says, 'if you row a boat from two ends, it can't go to both directions. It only goes towards one direction. I also want to follow one direction only – only Bangali'. On the one hand, Adi and Sabita's actions and desires of social distancing can be seen as a way of isolating themselves from the Bihari community; on the other hand, this also puts emphasis on their attempts to disassociate from being different and become part of the mainstream Bangladeshi community.

Unequal relationships of power

Everyday life experiences of children are embedded in the unequal relationships of power between children and adults, which are rooted in unequal gendered and generational power dynamics, that 'discriminate, exclude and violate' children in their everyday lives (Parkes 2015, p. 3). Building on the work of Alanen (2001) and Mayall (2002), Punch (2005, p. 169) argues that the relationships between children and adults are constructed on 'generationing' practices, which puts emphasis on exploring the practices of 'childing' and 'adulting' to understand how children and adults are constructed and positioned within a particular context. Shifting the attention from children's individual responses and actions to relational practices across generations helps to better understand the social and structural complexities and power dynamics within the society which affect children's lives (Ansell 2014).

At the outset, it is important to consider the gendered power dynamics in the camp, which set the tone to illustrate how this shapes understanding of power and authority among children and adults. In a group discussion, a few fathers made fun of the apparent 'powerlessness' of their wives as they said that the mothers lack control to discipline their children. They also boasted of their apparent authority, as while disciplining children, the mothers often threaten their children drawing references to their fathers. Children's submission to their fathers' authority and apparent indifference to their mothers' instructions reflect that often children, albeit unintentionally, contribute in reproducing the normative power dynamics in their relationships with adults. However, the patriarchal ideology was also evident in the way a few women expressed themselves about the practices of gendered relations and male domination in their families. Sohani, 28, justifies and normalises men's violence against women as she says that 'even the kings and monarchs abuse women, and we are mere poor people'. She believes that poverty increases chances for men to be more violent against women, as traditionally men carry the burden of looking after the family and the external challenges often strain their relationships.

Being raised in an environment where men's abuse towards women is normalised, and often trivialised, a number of male participants reported that they find it a threat to their patriarchal authority when women resist domestic violence. Ramjan, 45, told me that, while growing up, he observed that his mother used to remain silent in the face of his father's aggression. As he grew up, Ramjan carried with him the belief that women's 'silence' is the natural way of reacting to male violence. Thus, when their wives occasionally question or refuse to silently accept men's aggression, men find it a challenge to their normative practice of hegemonic masculinity. Men expressed their fears that their *maan-shomman* [respectability] might be jeopardised if women resisted their violence, as women's resistance breaks the power dynamics between men and women in the community. Similarly, women also expressed that they were afraid of losing their *maan-shomman* if their husbands beat them in front their children; as Koli said in a group discussion, 'isn't it *lojja* [shameful] to be beaten in front of the children? It is really shameful'. The drivers to lose *maan-shomman* might be different for men and women, however, they share a common threat where the normative gendered and generational power dynamics are challenged.

Parents' narratives presented later in this section illustrate that children, while exercising their agencies to confront violence, often immerse themselves in the process of (re)producing violence while challenging the power dynamics between children and adults (Holloway et al. 2018). The practices and beliefs on gendered power dynamics and inequalities lead both men and women to realise that their conflicting relationships of power might affect their inter-generational power in their relationships with their children. Shomota, 47, mother of Shopno, expressed her concern over an incident as she observed a son attempting to hit his mother. 'If a son attempts to hit his mum, then is there any *shomman* [respect] left for the parent?' Shomota continued, 'my son also held his dad's hand [when his dad wanted to smack him] that day'. Resonating with Shomota, Chapa, 30, also expressed her concern that her son once told her, 'if he [dad] beats you like this, I would beat him once I grow up'. Several parents shared the common fear that if children observe domestic violence, then they might fall into the cycle of reproducing violence. As many women prefer to remain silent to avoid further violence, they often teach their children to remain silent too. Parul, 40, said, 'my daughter says if anyone says a bad word to me then shouldn't I respond back. This is bad, really bad. I don't like violence at all. Allah will see what to do with the person who uses bad words. I will not say anything'.

Children's responses to violence

Mayall (2002, p. 46) argued that often children justify parental authority on the grounds that parents know more than their children and have 'a duty to protect them and provide for them'; thus children validated that parents have the right to make decisions concerning how children should lead their lives and how they should behave. Similarly, in their research with children living in the streets of Bangladesh, Conticini and Hulme (2007) demonstrated that children broadly frame violence from two perspectives, legitimate and illegitimate violence. The violence that is targeted towards children by the adults for disciplining and educating children is considered by children as *'nedgio bichar'* (fair punishment), and other violence which children could not justify was realised as *'onedgio bichar'* (unfair violence) (Conticini and Hulme 2007, p. 218). This resonates with my research too, in that I observed a few children who agreed that adults have the 'right' to hit children when children do something wrong, as they consider hitting as a tool in the learning process.

> Shimanto: Parents are *boro* [grown-ups]. We are *chhoto* [young]. We remain *nichu* [low] with them.
> Researcher: What is the reason for that?
> Shimanto: We must be afraid of our parents. Aren't they our parents? We must obey them.
> Mishuk: Nothing would happen even if they hit us.
> (Extracts from group discussion, boys' group)

Bablu, 10, in a group discussion, expressed his deep-down belief that hitting by the parents is acceptable and that it is an *'odhikar'* [right] of parents to hit their children. Bablu said: 'yes, they hit me. Why won't they? They have *odhikar* to hit our bodies'. Bablu justifies his parents' hitting towards him as their 'right' which is perceived as a 'legitimate adult reaction to a child's misbehavior' (Conticini and Hulme 2007, p. 218). Bablu reproduces the cycle of violence as he says, 'now you see, here [among the three other boys in that group] I am the oldest. Would it be a problem if I slap him [Bablu refers to Shurjo, a 9-year-old participant in the group]? There is no "fault". Now they [parents] hit me, they as adults can hit me. It doesn't matter at all'. As his parents have the right to hit their children, Bablu believes, he also has the 'right' to hit someone younger than him if they misbehave with him.

In contrast to absolute conformity to adults' authorities, children often demonstrated varied techniques to negotiate their position of power with adults. I present the case of Nishi to illustrate how children negotiate agency with adults as a way to stop violent practices, and what role family culture plays in mediating the process of negotiations. Nishi, a 16-year-old girl, is aware of the 'rotten' side of the camp, but believes that through the change in their attitudes, approaches and actions they can transform their lives into something positive. Nishi is unwilling to take it for granted that as she is from the camp, she would have to embrace its 'rotten' culture. Along with the generational practices in her cousins' neighbourhood, a few child rights sessions that she attended at a local NGO helped her to appreciate children's right to protection from violence. Thus, Nishi proactively engages in discussions with her parents where she strongly articulates her desires for a changed culture in their home, which she briefly summarised to me:

'If you [the father] scold or beat me – for no reason – one day I might throw this back to you. You would then say, how come that a daughter talks so badly with a dad! But it is you who started the *galagali* [yelling]'. Nishi continues, 'we have friends in school. They are wealthy and rich – even then they are my friends. If I ever get used to this *galagali*, then I might start using those with my friends too. They would then think that I have learned all these *galagali* as I grew up in the camp.'

However, Nishi's effort to negotiate agency with her parents was mediated by the parenting practices at their home which deviated from the dominant cultural norms of parenting at the camp. Rozina, 40, Nishi's mother, reinforced what Nishi told me earlier about parenting practices:

We live in the camp, but our family is different from others in the camp. Many, in this camp, fights inside their house, hit each other, we do not have these among us. We eat if we have food, otherwise, we starve. But we would never fight with others. Every family is not the same. There are many who do not have education, there are many do not interact with decent people. The environment, you know, the way people use their language tells about their ancestry. People do not understand this. This is the reason they engage in fights.

Their parenting values and practices created a safe space in their family for Nishi to have an open discussion with her parents about practices of violence and its impact.

Children's economic power and their contribution to the family often stimulate children to negotiate and create strategies to resist violence. In a group discussion, Rima, 12, said that she stopped going to the *karchupi* factory as the *mohajon* [the factory owner] smacked her if she did not do her work well. Rima told her mother and her older sister about the punishments and expressed her unwillingness to go to work any further. However, they were upset as they felt Rima stopped going to work for 'no apparent reason'. Rini, the older sister of Rima, told me that, 'we asked her so many times [about the reasons for her discontinuing work]! She didn't give a reason. She only says that *mohajon* scolds, swears and shouts. She stopped going to work only for this reason!' Even after Rima clearly expressed that she left work because of the punishments, Rima's sister and mother were reluctant to understand how such reasons could be valid 'excuses' to leave work. Hitting, smacking and shouting is so normalised that these are not considered as justifiable reasons to refuse to work. Nevertheless, Rima was able to remain firm on her position with the power of her economic influence in the family, as she has the potential for future income in the family.

Her mother and older sister did not want to upset her as even the little contribution she makes is valued by the family.

The power dynamics between parents and children are often negotiated through ideas of social capital (e.g. social network and connections) and cultural capital (e.g. education) (Morrow 1999). Adi, as I quoted earlier, is aware that if he does well academically, or behaves nicely with others, then he will be appreciated by people. Intentionally though, he only behaves well with those with whom he connects culturally. With others, who he considers are not 'educated' or 'cultured', he raises his voice and acts disrespectfully: 'I shout at them, yell at them, you know, I do a bit of *beyadobi* [disobedience] with them'. By his deliberate performance of disobedience, Adi attempts to set himself apart from most of the children in the camp. As a way of showing his nonconformity to his parents' authority, Adi told me that he does not talk to his parents about his academic progress or future plans. Adi said, 'they always remain tense about my studies and my future. I know that I am heading towards betterment. But I don't talk to them about this. I mean, they are, what should I say, they are sort of uncultured type. You can only explain to those who come from an educated background... they are only *chillachilli* [chaos and uproar]. They have chaos for everything'. With his education and connections with people outside the camp, Adi attempts to triumph over his parents' apparent lack of cultural resources and overturn aspects of the intergenerational power dynamics.

The findings of this research also show that children's birth order plays a crucial role in experiencing and responding to everyday violence. The case of Shiuly shows that she has to compromise her agency to create opportunities for her younger siblings. Shiuly, 20, is the second-born child among five siblings, and since their eldest sister moved into her in-law's house after getting married five years back, Shiuly has taken on the role of oldest sibling in the family. The uncertainty in their father's wages led Shiuly to sacrifice her education in order to take on *karchupi* work, which is the only source of steady income in the family. Apart from her financial contribution, Shiuly also embraced a 'motherly' role in the family, where her sisters feel the comfort and confidence to share their everyday problems with her in confidence.

> I keep all the problems to myself. I don't share this with my mom or my younger sisters. I listen to my younger sisters as they tell their problems. But I never share anything to my mother as she gets over stressed on every small issue. My mother is already overloaded with lots of family issues – poverty, laid off husband, reckless son, marriageable daughters, and what not! So I don't usually share our everyday issues with my mother. I take all the stresses to myself and relieve my mother from this additional stresses of us [siblings]. (Shiuly 20, female, group discussion)

Shiuly's dual roles to provide both economic and emotional support to the family, on the one hand create opportunities for her younger sisters to improve their academic prospects and wellbeing, while on the other hand make her overstressed and overloaded with burdens. Shiuly quintessentially demonstrates the fluidity of children's everyday experiences, where they are passive victim of violence on one hand, whereas on the other, they turn out to be the source of strength for themselves and for others. Shiuly looked fragile when she expressed her discontent that she always has to carry some '*chaap*' [stress] within her. However, a moment later, she seemed to be strong as she confidently said she possesses some '*shokti*' [inner strength] within herself which drives her to keep going and handle the hurdles of life with confidence.

Conclusion

There is little research available which explores the subtlety and nuances of children's experiences of everyday violence, particularly in the context of Bangladesh. While most literature focuses on either experiences of children living in an extremely violent context (Boyden and de Berry 2004, Hart 2008, Pells 2012, Seymour 2012, Markowska-Manista, this volume), or children's experiences of extreme forms of violence (Leonard 1996, Rockhill 2010), limited attention has been given to understand how childhood is realised in a socio-spatial context where inequality and deprivation are part of children's everyday lives. Bringing out the dominant discourses of childhood in a context which is often considered as 'rotten', and where childhood is realised through a normative binary of 'good child' and 'bad child', this chapter discussed children's experiences of violence as shaped by unequal gendered and generational relationships of power. Nevertheless, children's narratives demonstrated that parents' background and culture, children's own cultural capital and their economic authority rather than inherited resources from their parents, and their birth order, often shape the way they experience, respond to and resist everyday violence. The accounts of children and adults also illustrated that often children's experiences of violence are interconnected, where one form of violence triggers another. Given the complexity of the factors that contribute towards shaping children's experiences, it is important to have a flexible and situated understanding of children's lives to realise the nuances of children's experiences in the context of everyday violence.

Notes

1 In order to protect anonymity of research participants, pseudonyms are used for the camp, the NGO and all the research participants.
2 A saree is traditional attire for women, and the Biharis are skilled at weaving and designing fine Benarashi sarees which have their origin in Benaras (or Banaras or Varanas) in Uttar Pradesh in India
3 The term '*Maowra*' originates from the Maurya Empire. The present day Bihar was part of the Maurya Empire in the time of ancient India around 300 BC. Now the term '*Maowra*' is used abusively by some Bangalis to refer to the 'non-Bengalis' and 'Urdu-speaking Bihari' people.

4 The term '*rajakar*' means collaborator, and is used to denote the war criminals who collaborated with Pakistan during the liberation war of Bangladesh in 1971.

References

Alanen, L., 2001. Explorations in generational analysis. In: L. Alanen and B. Mayall, eds. *Conceptualizing Child-Adult Relations*. London: Routledge.

Ansell, N., 2009. Childhood and the politics of scale: Descaling children's geographies? *Progress in Human Geography*, 33(2), 190–209.

Ansell, N., 2014. Commentary: 'Generationing' development. *The European Journal of Development Research*, 26(S2), 283–291.

Atkinson-Sheppard, S., 2017. Street children and 'protective agency': Exploring young people's involvement in organised crime in Dhaka, Bangladesh. *Childhood*, 24(3), 416–429.

Balagopalan, S., 2014. *Inhabiting 'Childhood': Children, Labour and Schooling in Postcolonial India*. New York: Palgrave Macmillan.

Bartlett, S., 2018. *Children and the Geography of Violence: Why Space and Place Matter*. Oxon: Routledge.

Basu, I., Devine, J., and Wood, G., 2018. *Politics and Governance in Bangladesh*. Oxon: Routledge.

Bourdieu, P., 2004. Gender and symbolic violence. In: N. Scheper-Hughes and P. Bourgois, eds. *Violence in War and Peace: An Anthology*. Malden: Blackwell Publishing Ltd., 339–342.

Boyden, J. and de Berry, J., 2004. *Children and Youth on the Front Line: Ethnography, Armed Conflict and Displacement*. Oxford: Berghahn Books.

Chakraborty, K., 2009. 'The good muslim girl': Conducting qualitative participatory research to understand the lives of young muslim women in the bustees of Kolkata. *Children's Geographies*, 7(4), 421–434.

Conticini, A. and Hulme, D., 2007. Escaping violence, seeking freedom: Why children in Bangladesh migrate to the street. *Development and Change*, 38(2), 201–227.

Dunkley, C.M. and Panelli, R., 2007. 'Preppy-jocks', 'rednecks', 'stoners', and 'scum': Power and youth social groups in rural Vermont. In: R. Panelli, S. Punch, and E. Robson, eds. *Global Perspectives on Rural Childhood and Youth: Young Rural Lives*. New York: Routledge, 165–177.

Farmer, P., 2004. An anthropology of structural violence. *Current Anthropology*, 45(3), 305–325.

Galtung, J., 1969. Violence, peace, and peace research. *Journal of Peace Research*, 6(3), 167–191.

Hart, J., 2008. *Years of Conflict: Adolescence, Political Violence and Displacement*. Oxford: Berghahn Books.

Holloway, S.L., Holt, L., and Mills, S., 2018. Questions of agency: Capacity, subjectivity, spatiality and temporality. *Progress in Human Geography*. doi:10.1177/0309132518757654

Holloway, S.L. and Valentine, G., 2004. *Children's Geographies: Playing, Living, Learning*. London: Routledge.

Hopkins, P. and Pain, R., 2007. Geographies of age: Thinking relationally. *Area*, 39(3), 287–294.

James, A., Jenks, C., and Prout, A., 1998. *Theorizing Childhood*. Cambridge: Polity Press.

Kraftl, P., Horton, J., and Tucker, F., 2012. *Critical Geographies of Childhood and Youth: Contemporary Policy and Practice.* Bristol: The Policy Press.

Leonard, L., 1996. Female circumcision in Southern Chad: Origins, meaning, and current practice. *Social Science and Medicine*, 43(2), 255–263.

Matthews, H. and Limb, M., 1999. Defining an agenda for the geography of children: Review and prospect. *Progress in Human Geography*, 23(1), 61–90.

Mayall, B., 2002. *Towards a Sociology for Childhood: Thinking from Children's Lives.* Buckingham: Open University Press.

Mayall, B., 2012. An afterword: Some reflections on a seminar series. *Children's Geographies*, 10(3), 347–355.

Montgomery, H., 2009. *An Introduction to Childhood: Anthropological Perspectives on Children's Lives.* London: Wiley Blackwell.

Morrow, V., 1999. Conceptualising social capital in relation to the well-being of children and young people: A critical review. *The Sociological Review*, 47(4), 744–765.

Parkes, J., 2007. The multiple meanings of violence: Children's talk about life in a South African neighbourhood. *Childhood*, 14(4), 401–414.

Parkes, J., 2015. Introduction. In: J. Parkes, ed. *Gender Violence in Poverty Contexts: The Educational Challenge.* London: Routledge, 3–10.

Paulsen, E., 2006. The citizenship status of the Urdu-speakers/Biharis in Bangladesh. *Refugee Survey Quarterly*, 25(3), 54–69.

Pells, K., 2012. 'Rights are everything we don't have': Clashing conceptions of vulnerability and agency in the daily lives of Rwandan children and youth. *Children's Geographies*, 10(4), 427–440.

Pells, K. and Morrow, V., 2018. *Children's Experiences of Violence: Evidence from the Young Lives Study in Ethiopia, India, Peru and Vietnam.* Oxford: Young Lives.

Punch, S., 2005. The generationing of power: A comparison of child-parent and sibling relations in Scotland. *Sociological Studies of Children and Youth*, 10(March), 169–188.

Punch, S., 2007. Generational power relations in rural Bolivia. In: R. Panelli, S. Punch, and E. Robson, eds. *Global Perspectives on Rural Childhood and Youth: Young Rural Lives.* London: Routledge, 149–164.

Redclift, V., 2010. Conceiving collectivity: The Urdu-speaking Bihari minority and the absence of home. In: L. Depretto, G. Macri, and C. Wong, eds. *Diasporas: Revisiting and Discovering.* Oxford: Inter-Disciplinary.Net, 311–327.

Rockhill, E.K., 2010. *Lost to the State: Family Discontinuity, Social Orphanhood and Residential Care in the Russian Far East.* Oxford: Berghahn Books.

Scheper-Hughes, N. and Bourgois, P., 2004. Introduction: Making sense of violence. In: N. Scheper-Hughes and P. Bourgois, eds. *Violence in War and Peace: An Anthology.* Malden: Blackwell Publishing Ltd, 1–31.

Seymour, C., 2012. Ambiguous agencies: Coping and survival in eastern Democratic Republic of Congo. *Children's Geographies*, 10(4), 373–384.

Wells, K. and Montgomery, H., 2014. Everyday violence and social recognition. In: K. Wells, E. Burman, H. Montgomery, and A. Watson, eds. *Childhood, Youth and Violence in Global Contexts: Research and Practice in Dialogue.* Basingtoke: Palgrave Macmillan, 1–15.

Winton, A., 2005. Youth, gangs and violence: Analysing the social and spatial mobility of young people in Guatemala City. *Children's Geographies*, 3(2), 167–184.

12 Discovering difference in outer suburbia

Mapping, intra-activity and alternative directedness in Shaun Tan's *Eric*

Amy Mulvenna

Chapter summary

This chapter follows Eric, a little 'alien' foreign exchange student-come-geographer-come-spatial-anthropologist, through a literary-spatial critique of the tale by the same name, taken from Shaun Tan's (2009) *Tales from Outer Suburbia*. Focusing on the materialities of mapping and the materiality of embodied place-making within 'Outer Suburbia', and looking to dimensions *beyond* the literal for significance, it is my aim to critique and disseminate Shaun Tan's work towards new audiences, positioning Shaun Tan as a major figure within contemporary childhood studies, children and youth geographies, and beyond. Moving from the pantry to the pavement and back again, I explore othering as a *process* of alternative directedness that implicates children's everyday politics; specifically, the agencies and politics of different materialities of the everyday, which are less than overt and obvious. In doing so, I discuss ways in which the counterpoint between pictures and words initiates young readers into a cartographic language that invites them to identify and critique attitudes towards otherness, strangeness and difference within the stories as well as real-life contexts. Ultimately, this chapter underscores the formative potential of *Eric* as an innovative and creative approach to teaching geography within the primary classroom *as well as* doing children and youth geographies *creatively*.

Introduction

> Art and literature particularly allow us to see behind the surface of material things and events because they transfigure the lives of the city and its inhabitants by means of their idiosyncratic, creative power of the imagination. (Sebastian Groes, 2011)

Groes suggests that art and literature can offer nuanced and profound criticalities through which we gain deeper understandings. It is entirely legitimate, I argue, that young adult and children's literature should enrich and vivify methodological approaches and debates within children and youth geographies and vice versa, building on Tara Woodyer and Hilary Geoghegan's (2013) provocative celebration of enchantment within human geography. In this chapter, I consider a tale by Australian author-illustrator Shaun Tan that deals imaginatively with an 'othered' young subject who negotiates and plays with place, space and *things* within an unfamiliar suburban environment. In my reading of the text, I explore how Tan brings mapping, touch and materiality together metaphorically, imaginatively, *and most intimately*, through the character of the eponymous miniature hero, Eric. I examine ways in which Eric's intra-actions *with* and plotting *of* emergent im/materialities provide starting points for deviations from known routes. Eric's mapping of suburbia quietly excites a process of what I call 'alternative directedness', inviting readers to identify, explore and critique attitudes towards strangeness and difference within the story as well as real-life contexts. I begin by outlining the tale, discussing how suburbia and its material culture emerge and transform simultaneously, going off in myriad directions. In the latter half of this chapter, I share ways in which *Eric* might be used practically in place-based research with children and young people, foregrounding crossovers between academic and primary geography.

What's in a name?

Meet our eponymous hero, Eric. His story is one of fifteen vignettes from Tan's illustrated picture book *Tales from Outer Suburbia*, published in 2009. Of all the tales, only *Eric* was subsequently published in 2010 as a stand-alone picture book, granting literary independence to a most fascinating little artist, geographer and cartographer. From the beginning, Eric is *othered* by the way he is framed in the tale. He is presented as a visitor, a 'foreign exchange student' (Tan, 2009, p.8). We are not told how old he is, nor where he is from. Although the family tries to make him comfortable, Eric is content to sleep and study in the pantry: '"It must be a cultural thing", said Mum' (*ibid.*, p.9), a phrase that becomes something of a refrain for difference throughout the tale.

Eric is therefore geographically and culturally displaced from his home. The host family further iterate this displacement: 'we found his name hard to pronounce, but he just told us to call him Eric'. Eric becomes 'Other-wise' in the noun form he is given by the family, his identity signed and qualified by virtue of being other. And yet, Tan foregrounds Eric's othering not to depict processes or struggles associated with 'fitting in'; rather, difference emerges more subtly in this tale – *in posse* – where Tan quietly relays the transformative potential of materiality as a creative and embodied process of place mapping. This in turn enables the reader to perceive *things* anew. For it is not only the material culture of suburbia that is invoked and performed, but also what Latham and McCormack (2004, p.705) would describe as the 'excessive potential of the immaterial', as I go on to explain.

Beyond the site of the home is 'Outer Suburbia' and the space of the street. The grey suburban world depicted in these frames appears to be resolutely

Figure 12.1 Eric arrives. © Shaun Tan's 'Eric'. Image reproduced with permission from *Tales from Outer Suburbia* by Shaun Tan, Templar Publishing, London.

'human', and much larger than the miniature hero. Having made lots of plans for exciting excursions with the new visitor, the child narrator expresses his perplexity at Eric, who instead is orientated towards the ostensibly banal little things he finds on the pavement, noticing their patterning and fine-grained details. His curiosity is focused on everyday household objects and appliances like sink plugs, postage stamps, bottle tops, cereal boxes and sweet wrappers; i.e. all the incidentals, the detritus or excesses of a habitus recognisable to many readers.

The tiles in Figs. 12.2 and 12.3 exemplify narrative economy on Tan's part: framings that give intimate yet disparate windows onto the familiarity of suburban lives. They are, furthermore, 'of the moment' – evocative of an aesthetic of the everyday as framed by polaroid shots.

In many of the best fairy tales, there are the ghosts that lurk behind the wanderer in the woods. Shadowing Eric at all times is a Benjaminesque reading of the sub/urban. More specifically, it is Benjamin's (2002) liminal figure of the ragpicker that emerges from between the 'cracks' of these discrete, crepuscular tiled illustrations. To explain: Benjamin's (2002) materialist analysis in *Das Passagen-Werk* (*The Arcades Project* – posthumously assembled notes that comprised a 'literary montage' [2002, p.460]) sought to show the 'refuse' or 'leftovers' of history; that is, the excluded or less than obvious

Figure 12.2 Questioning at home. © Shaun Tan's 'Eric'. Image reproduced with permission from *Tales from Outer Suburbia* by Shaun Tan, Templar Publishing, London.

amidst the ready-made. He compared the ragpicker to the poet, likening their efforts to collect, archive and *feel* their way through waste materials – literal and literary – that are in turn re-assembled to create or perform *other* possibilities, *other* (his)stories (Le Roy, 2017). Just as the ragpicker demonstrates acute attention to detail, Eric handles a host of curiosities exploringly, marking his journeying in and through Outer Suburbia by coming into direct contact with that which he encounters on the ground through his touch. Eric does not simply *see*; rather, he attends to, notices, gestures, touches, and senses all within the intimate and everyday spacings through which he moves. And he does so wordlessly, which is part of the fascination of the tale: Tan ranks the haptic on a par with the senses of vision and hearing, which, as Tim Ingold points out, is traditionally less common within Western modes of thought: '...ever since Plato and Aristotle, the western tradition has consistently ranked the senses of vision and hearing over the contact sense of touch' (Ingold, 2004, p.330). Erin Manning explains the significance of touch for bodies-in-action, remarking on potentiality for responsiveness, relations and inventions:

> I touch not by accident but with a determination to feel you, to reach you, to be affected by you. Touch implies a transitive verb, it implies that

Figure 12.3 Exploring the street. © Shaun Tan's 'Eric'. Image reproduced with per-
    mission from *Tales from Outer Suburbia* by Shaun Tan, Templar Pub-
    lishing, London.

> I can, that I will reach toward you and allow the texture of your body to
> make an imprint on mine. Touch produces an event. (Manning, 2007,
> p.12)

Through Eric's pedestrian encounters, the materialities and temporalities of
place exceed the purely cognitive and semiotic. By pausing to touch and
attend to the heterogeneous material reality of suburbia around him, Eric
foregrounds the unexpected and inexplicable; an entanglement of *other* ways
of knowing and sensing the 'more-than' of place (Lorimer, 2005; Thrift,
2008) – an alternative directedness as it were. Indeed, this more fluid way of
knowing and going is suggested by the artefacts themselves, many of which
connote markers of passage or movement through, around and between
spaces / places: I think of the flow of water denoted by the plug hole, the
travelling mailbags and letters suggested by the postage stamp.

Related to this theme of travel, Eric's little suitcase (a peanut) is important,
for it is the kernel that gave birth to this allegory of difference as it were. In an
interview with fellow author Neil Gaiman (2011), Tan explains: 'It all started
with looking at nuts and drawing them and thinking wouldn't it be great to
have an entire culture, a civilisation, based entirely around nuts and the sheer

variety of different types of nut.' Suburban travel is further represented in the comedic image of Eric strapped in by his comically oversized buckle, his legs and feet out of action (Fig. 12.4).

Being thus restricted, Eric is ensconced within a familiar trope of suburbia. 'Destination-oriented travel' (to borrow Ingold's phrase) was popularised at the same time as the growth of post-war city suburbs, when families could afford their own cars for the first time. As Ingold explains, this form of travel:

> ...encouraged the belief that knowledge is built up not along paths of pedestrian movement but through the accumulation of observations taken from successive points of rest... But in real life... we do not perceive things from a single vantage point, but rather by walking around them... Cognition should not be *set off* from locomotion, along the lines of a division between head and heels, since walking is itself a form of circu-mambulatory knowing. (Ingold, 2004, p.331)

To the relationship between our eye and our intellect (following Ingold's reasoning) we should add the foot, for movement is paramount when it comes to

Figure 12.4 In locomotion. © Shaun Tan's 'Eric'. Image reproduced with permission from *Tales from Outer Suburbia* by Shaun Tan, Templar Publishing, London.

perceiving things in our everyday lives. Ingold stresses the essential link between *perception* and locomotion in his earlier writing:

> ...if perception is thus a function of movement, then what we perceive must, at least in part, depend on how we move. Locomotion, not cognition, must be the starting point for the study of perceptual activity. (Ingold, 2004, p.166)

Cognitive scientist Robert Solso notes how our minds add interpretation to our perceptions: everyone 'see[s] the world in profoundly different ways because of the vast diversity in the way we humans develop individual mental structures of the world' (Solso, 1996, p.3). Hence, the act of perception encompasses two standpoints: on one hand it is an 'embodied experience', familiar to every human being (Crossley, 1995, p.45). And yet, on the other hand, perception is also a highly unique and personal bond through which the 'perceiver [as an individual] and the perceived become relational beings' (*ibid.*, p.46). The second viewpoint resonates with Eric's enquiry (and encounters), characterised by 'a fundamental orientation towards the ground', described in Japanese by the word 'Kawada' (Ingold, 2011, p.330). Eric's orientation is both recuperative and generative, too, as we later discover in the final double-page spread, insofar as it foregrounds uncertainty, openness – even confusion – a rhizomatic reading and mapping of place that has no beginning or end, 'but always a middle (milieu) from which it grows and overspills.' (Deleuze & Guattari, 1987, p.21). Yet before discovering the end gift, readers must further orientate themselves to a series of curious pedestrian-events within Outer Suburbia.

Eric's encounters displace any preconceptions of suburban spacings: the size and arrangement of these crepuscular illustrations do not easily allow us to create a mental image of the scale or scope of the host family's suburbia. Instead, we were brought to focus on the detail within these intimately arranged tiles, all of which are scaled to his small stature (Fig. 12.5).

Our 'reading' of Outer Suburbia is therefore akin to how literary critic Eric Bulsen describes the process of 'reading the city' in works by modernist authors like Joyce and Woolf (Bulsen, 2007, p.110). Bulsen compares this reading to a delayed process of piecing together an 'urban blueprint'; that is, a kind of plan or map of the city. In reading *Eric* as something of an alternative roadmap, we must piece together the disparately connected monochrome tiles, each one interrupted by the white space of the guttering between the intimately arranged panels – suggestive perhaps of the social isolation and alienation shared by inhabitants of suburbia (*ibid.*, p.112). This reading chimes with Ingold's lament of human detachment from urban streets, where he describes the 'groundlessness of modern society, characterised by the reduction of pedestrian experience to the operation of a stepping machine' (Ingold, 2004, p.329). He continues pessimistically:

Figure 12.5 In the pantry. © Shaun Tan's 'Eric'. Image reproduced with permission from *Tales from Outer Suburbia* by Shaun Tan, Templar Publishing, London.

It appears that people, in their daily lives, merely skim the surface of a world that has been previously mapped out and constructed for them to occupy, rather than contributing through their movements to its on-going formation. (*ibid.*)

It follows that we as readers lack any real sense of orientation – at least, that is, until the very end when the host family discover the gifts left by Eric inside the pantry. Indeed, his legacy is revealed only after his departure, floating through a window on a leaf (Fig. 12.6), which emphasises a mood that can be described as liminal: he has moved on from the pre-ritual state, but it is we the reader who must complete the ritual by entering the pantry once more.

The future is ...in the pantry

We return finally to the intimate spacing of the pantry (where Eric chooses to sleep and study) to discover an immersive and dizzying abundance of organic growth that persists across the double page spread (Fig. 12.7). Readers are invited by the narrator to 'go and see for yourself. It's still there after all these years...' (Tan, 2009, p.17). In responding to this call by turning to the final pages of the book, the reader reveals the vibrant outcome(s) of a process that

Figure 12.6 Eric's departure. © Shaun Tan's 'Eric'. Image reproduced with permission from *Tales from Outer Suburbia* by Shaun Tan, Templar Publishing, London.

could not be expressed in words. There is no obvious temporal, spatial or chronological order to the creative liveliness, richness and density; rather, the domestic scene visualises light and growth flowing freely forth from formerly discarded things, unrestricted by text, dissolving the idea that they are simply disposable, and furthermore blending the various material and socio-cultural spacings from whence they came. Exposed now in light and colour, this alternative 'urban blueprint' (a continuous dialectic) showcases enchantment within the banal or low key, in that which is typically not considered to hold any significance: the 'sub' urban as it were.

We realise now that through his exploratory wayfaring – a coupling of not only 'locomotion and perception' (Ingold, 2007, p.78) but also encounter and intra-action, merging movement, gesture, vision, touch and *things* most intimately – Eric has responded to and expanded upon different material features (pattern, detail, position and perspective) of the stamps, bottle tops, and sweet wrappers. These features are connected to different ways of sensing, exploring, moving through, and engaging with place as agentic assemblage (Duhn, 2012) through *child-led* practices of walking, collecting, recording, curating, and wondering (Hamm, 2015). That which was initially seen to be familiar now houses the unfamiliar, where Eric recycles artefacts *beyond* the echoes of Fordism and mass produced stuff(s), countering the idea that suburbia is to be

Figure 12.7 The gift. © Shaun Tan's 'Eric'. Image reproduced with permission from *Tales from Outer Suburbia* by Shaun Tan, Templar Publishing, London.

taken as given and familiar. Eric's wayfaring thus entangles sensorial, situated and imaginary emplaced knowledges in ways that can be very different from top-down, policy-driven objectives for place.

To give some context: the 1950s signalled the birth of suburbia in the United States, UK, and Australia. Initially developed as a 'bulwark against communism and class conflict' (that is, threats of the 'other'), 'suburbia' was upheld by strict socio-cultural norms and values (May, 1989, p.158). Contrastingly, I argue that Tan's 'Outer Suburbia' is metonymic for alterity: a *terra incognita* open to mystery, desire, fancy – monstrosity even – that would counter or threaten an excessively ordered world. From the title alone, Tan himself emerges as migrant thinker, wondering and wandering as he puts it 'at the edge of consciousness'.

> 'Outer Suburbia' might refer to both a state of mind as well as a place: somewhere close and familiar but also on the edge of consciousness (and not unlike 'outer space'). Suburbia is often represented as a banal, quotidian, even boring place that escapes much notice. Yet I also think it is a fine substitute for the medieval forests of fairytale lore, a place of subconscious imaginings. (Tan, 2014)

Tan grew up as the son of Chinese migrants in 1970s Perth, Australia. It was here he tells us that, 'my parents pegged a spot in a freshly minted northern suburb that was quite devoid of any clear cultural identity or history' (*ibid.*). I want to stress just how deeply Tan seems to have been influenced by the geophilosophy of this 'universe of bulldozed "tabula rasa" of coastal dunes, and fast-tracked, walled-in housing estates', and the way in which the peculiarity of the region has effected an imaginative place mapping in his mind. For as Tan's recollections of childhood stir his mind to wonder *beyond* the banality and quotidian normalcy of suburban Perth's 'empty footpaths, shady parks, rows of blank-faced houses, deeply shadowed windows and wide roads', he recognises latent possibilities within the suburbs; entryways and exits suggest multiple, unfolding, even mythical dimensions within many of his tales.

Following Noora Pyyry's (2017) thinking, I recognise that dwelling *with* the sub/urban can be monstrously alluring and enchanting; a mode of fast-moving, excessive, and wholly affective becoming – an experience nigh on impossible to put into words. Such is the concept of 'monstrosity' itself: a fluid and ambiguous term that can be stretched in myriad different ways, exceeding fixed representation, unity, and meaning, which attests to the sheer imaginative force behind it. To look at it another way, adopting a Marxist post-phenomenological perspective, a key emphasis in this tale is that things and materials are processual and open to heterogeneous possibilities of becoming otherwise, particularly where questioning of the local as sites of wonder and curiosity is actively encouraged – in this case, by a very little 'monster'.

Eric, being other-*wise* himself – an active and profound producer of knowledge – gifts the host family and reader with an*other* way of reading suburbia, disavowing predictability and tidiness. Rather, he displays an irradiated openness to difference that continues to unfold in the temporal and spatial heterotopia of the pantry. To be sure: this microcosm of budding growths is Eric's gift to this place; that is, the gift of attention *to* the 'material elements of place', as Sacks and Zumdick would put it (2013, p.56). Moreover, as I see it, this gift, emblematic of place wisdom, can be read through the lens of those participatory, place-based research methodologies as discussed by Karen Malone (2018), Professor of Education and Sustainability Research at Western Sydney University.

Reflections on difference, materiality, place and situatedness are keystones to Malone's seminal text, and similarly underpin the creative pedagogical inquiry of Dr Helen Clarke and Sharon Witt, Primary Geography educators at the University of Winchester. They read Eric's place wisdom – a form of being and knowing – as a directive for educators and scholars to think about children's entanglement in 'real world' geographies. Together, they respond to the miniature hero through their collaboration, Attention2Place (A2P), seeking to facilitate children's responses *to* and intra-actions *with* bodies, things and materials within their local geographies. Their inquiry intersects everyday, local geographies with lyrical responses to place and place-relations. This approach stems from their belief that sensory immersion through haptic encounters is key to encouraging children to become 'research active': attending to, questioning, feeling, touching, following, plotting and crucially, storying.

I discuss Clarke and Witt's work in more detail in the following section. More generally, however, their approaches to place-making are conceptually situated within an international collective of childhood and feminist scholars and educators invested in a 'common worlds' approach (a term borrowed from Bruno Latour [2004]). This approach recognises the indivisibility of 'nature' and the 'social', and attends to human relations with other things and species. Common Worlds Research Collective co-founder, Affrica Taylor (2017, pp.11–12), describes one strand of this approach as exploring ways in which children and young people make, disorder and vitally remake place through entangled 'worldly relations'. Indeed, we can read Eric's literal and lyrical pavement-to-pantry enquiry as akin to the countless micro-cultures that children continually progress in and out of as they explore and engage (both multisensually and multimodally) with different types of spacings, performing their own temporal, material and processual rituals, habits and customs (Horton & Kraftl, 2006a; 2006b).

Over the past ten years in particular, new materialist and posthumanist ontologies and methodologies have been fruitfully brought to bear on research regarding children and young people's (tacit) place-based knowledges, wisdoms and practices *across* disciplines in both the global north and south (Horton & Kraftl, 2017; Malone, 2016; Nxumalo, in press; Pacini-Ketchabaw & Clark, 2014; Pacini-Ketchabaw & Taylor, 2015; Yazbeck & Danis, 2015). Research within and beyond the Common Worlds Collective advocates relational approaches to place-based research with children and young people as a means to nurture creativity, place-responsiveness and place wisdom as active endeavours. Conceptually advancing theories of assemblage, encounter and affect, members of the Collective affirm difference and change as immanent in even the most minute ways, in the sense that Waterton and Dittmer (2014, p.123) describe where they draw attention to micro 'ways in which the social and material interact in ongoing processes of reconfiguring the world'. For example, starting from Karen Barad's (2003) concepts of 'thing-power' and 'intra-action', Pauliina Rautio's (2013) work exemplifies the researcher's responsiveness to momentary encounters, emphasising ways in which children affect and are affected by the stones they carry in their pockets. Reflecting on her three year old son's responses to the materiality of the urban environment ('thinking with broken glass'), Noora Pyyry (2017) likewise acknowledges the contingent continuum between children's playing, unmaking, re/making and doing and their intra-actions with *other* bodies, things and materials.

With all of this in mind, to read *Eric* as a simple narrative of 'foreigner comes to stay with family and learns about their culture' is to fail to acknowledge and explore the nuances of this allegory of difference. In his *queering* of suburbia as it were, Eric's tale advances an expanded field of enquiry, encouraging attention to place – place that is comprised of the flows of myriad *other* bodies, both animate and inanimate that are agentic also, and continually cross-cut the lives of children and young people.

What is more, I argue that *Eric* is ripe for geographical and artful approaches to enquiry with children and young people that researchers invested in creative and alternative methodologies would do well to attend to. In the following section, I discuss a selection of practical approaches to place-responsive research with children, recognising that knowledge of place is emergent, embedded and embodied, material *and* immaterial, insofar as both aspects necessarily 'contribute to social production' (Fox & Alldred, 2015, p.399).

Eric-as-methodological approach, or doing children's geographies *creatively*

Intrepid traveller that he is, Eric has been causing quite a stir in real-world geographical circles of late. Clarke and Witt accompanied Eric to the 2018 Geographical Association (GA) Annual Conference and Exhibition at the University of Sheffield. Eric has reportedly also been spotted by numerous children in fieldwork spaces across Belfast (most recently in the baked beans aisle in Tesco's, East Belfast). It is my aim now to draw on creative (methodological) approaches from school-based and academic geographies; specifically, Clarke and Witt's A2P collaboration and my own PhD fieldwork with children in Belfast, aged 7–11. I do so with three objectives in mind. Firstly, to share practical ideas on how *Eric* might be deployed empirically, with a view to support children and young people's engagement *with*, explorations *of*, and responses *to* real world geographies (both human and physical), as well as imagined geographies / geographies of the imagination. The suggested activities are variously designed to mobilise participants' curiosity, open-ended thinking, wondering, making and doing, as rooted in concepts of: place, temporality, materiality, local sites, spacings and situations, as well as the 'more than' of place. Secondly, in keeping with the themes of this edited collection, I draw on nascent reflections from my fieldwork in order to consider some ways in which Eric has allowed me to (re)consider the affective and material within geographies of division in Belfast. Lastly, and more broadly, I advocate the use of picture-book-as-method in children and youth geographies. It has been over a decade since Horton and Kraftl (2006a, p.69) pronounced 'children's geographies could do more'; and yet the potential of story and storytelling as a mode of (collaborative) inquiry and co-construction of meaning is very much under-explored to date. By engaging with story-centred and / or picture-book-orientated methodologies, researchers might very well do more, by creatively deploying story – i.e. that which is very real and resonant in the lives of many children – to *listen back to* participants in ways that facilitate reciprocal, respectful and inspiring relationships between children, researchers, books and the world.

'Geo Findspots' and Eric

Clarke and Witt (2018) advocate a four-stage, intra-active pedagogical approach, using *Eric* as stimulus for place-based, child-led, creative enquiry. Their framework intersects National Curriculum requirements (within the

UK) with spatial theories orientated towards new materialist and post-humanist philosophies. These range from assemblage thinking (assemblage of place, assemblage as complexity of place, and assemblage as representation of place) to theories of creative encounter. Underpinning each stage is a neat archaeological hook. Clarke and Witt (2018) introduced Eric to audience members at the GA conference in Sheffield along with the concept of the 'findspot' – that is, the place or location where an object / artefact is found. Together, they have developed the thought-provoking concept of the 'Geo Findspot': '...a place where a range of objects can be found. It invites the finder to build a picture of the place' (Clarke & Witt, 2018). Openness to enchantment is key: they describe locating Geo Findspots as akin to 'finding a little bit of magic in the everyday' (*ibid.*). Their aim: to incite tacit, emplaced knowledge; to motivate and mobilise mindful and active enquiry *beyond* the classroom walls, encouraging children to attend to the particularities and specificities of the field and everyday geographies. Geo Findspots can include: 'locational geography, human / physical features, site / situation, seasonality and environmental change, assemblage of materials and sense of place' (*ibid.*). Figs. 12.8a–d showcase just some of Eric's wanderings and wonderings to date.

Finding

As a starting point, Clarke and Witt recommend reading the story before moving on to questioning in order to elicit explicit engagement with locational knowledge. Children are encouraged to consider and reflect on both human and physical features of places around them. Emphasis is on *finding out* using one's local expertise.

Enquiring

Stage two involves active enquiry through place-responsive fieldwork: a fieldwork of attentiveness that necessitates paying attention to and dwelling *with* place while moving through it, unrestricted by supposed 'parameters' of the literal and lyrical. Clarke and Witt suggest that attention to the 'more than' of place might be mobilised multimodally, including: photography, mappings, interviews, field sketches and more, which in turn may invoke tacit or unexpected spatial knowledges. This stage of enquiry is underscored by receptivity to enchantment, reminiscent of Robert Macfarlane's writing on children's apprehension of their environments. He is worth quoting at length here:

> To young children, of course, nature is full of doors – is nothing but doors, really – and they swing open at every step. [...] To a three- or four-year-old, 'landscape' is not backdrop or wallpaper, it is a medium, teeming with opportunity and volatile in its textures. [...] What we bloodlessly call 'place' is to young children a wild compound of dream, spell and substance: place is somewhere they are always in, never on. (Macfarlane, 2015, p.315)

Figure 12.8 Discovering Geo Findspots with Eric! © Dr Helen Clarke and Sharon
Witt. Image reproduced with permission from the authors.

Responding

Stage three involves facilitation of children's responses to and recordings
in, of and about place. The mutability and complexity of the literal
(found objects) and lyrical (imagined realms) are contingent here, where
the empirical focus turns towards collecting, collating, curating, and

Figure 12.8 (continued)

crucially *celebrating* place and emplaced knowledge. Indeed, artfulness and creativity abound where the geographical imagination is excited! With an eye to inclusivity, Clarke and Witt offer various strategic modes for engagement with local places as sites of curiosity and wonder: through collage, story, poetry, movement and cabinets of curiosity. A

Figure 12.8 (continued)

particular favourite of mine is their use of 'Geo Findspot boxes' (inspired by Cornell Boxes, or 'dream boxes'). These acts of creation and curation explore associations that cannot (easily) be grasped or expressed in words, i.e. memories, feelings, and emotions that may intersect the everyday and the extraordinary.

Figure 12.8 (continued)

Questioning

During the fourth and final stage, children are invited to reflect and *question* those embodied and sensory experiences from the field; an affirmation of sense-making and emplaced wisdom as it were, which informs their personal geographies going forwards. Creativity abounds in Clarke and Witt's model, which is a similar point of departure for my own methodology, focusing specifically on place mapping, as I now outline in brief.

Mapping with Eric in Belfast: a flavour of my creative cuts

Cultural differences are entrenched and performed within Belfast, Northern Ireland, still today, 20 years since the signing of the Belfast Agreement / Good Friday Agreement. Responding to Peter Shirlow's (2006, p.106) argument that, 'the fundamental problem that affects Belfast is that geography matters in ways that are overt and obvious', my research with children aged 7–11 across Belfast focuses on creative mapping as an alternative methodology. Working participatorily with over 100 children aged 7–11 in primary schools and youth centres, I have facilitated workshops around children's place-knowing and place-mapping, focusing on *less-than* obvious, *less-than* overt, situated, affective, and embodied everyday geographies and place-making practices. I do so with the aims of enlisting participants creativity and potentiality, to unfold different ways of framing and thinking about Belfast from children's perspectives, and to further explore what these creative mappings and encounters might tell us about geographies of division in Northern Ireland today.

I now discuss in short just two of our creative cuts to give a flavour of what we have been getting up to with Eric during our multi-modal data-making. I use the phrase 'creative cuts' as opposed to methods, inspired by material feminist thinking and Karen Barad's (2003, 2011) use of the phrase 'agentic cuts'. Barad's use of the term accords with her 'agential realist ontology', which understands human existence as co-constituted through processes of 'intra-action' with other bodies, human and non-human (2003, p.822). The concept of the 'cut' or 'cutting' in materialist research-creation concerns the 'entanglement' of bodies (Ingold, 2008): where bodies encounter and intra-act they cross-cut each other's dynamic materiality, signalling acts of 'things coming-together apart *in one movement*…engender[ing] a discontinuous passage where something new emerges' (Springaay & Zaliwska, 2015, p.141). To be sure, the creative cuts are not about amassing or capturing 'data' (Boyd, 2017; Somerville, 2015, 2016). Rather, the thinking behind them was that the related workshops would bring participants into encounters with the materiality and the *more-than* of the field site and their local geographies: for example, through story, movement, crafting and moulding. Understanding the children's creativity (and my own) as emerging from ongoing, experimental, experiential processes between 'makers' (the children and myself), materials, tools, and other non-human things, these mapping workshops were all about mingling *with* and emerging *from* place as comprised of the flows of countless *other* bodies, both animate and inanimate that are agentic also (Bennett, 2010; Hultman & Taguchi, 2010), and continually cross-cut the lives of children and young people.

Outlining the creative cuts

Picnicking

Eric's Belfast journeying began in a context not so far removed from his pantry dwelling place. Responding to the popularity of the tuck shop in all field sites, I intuited that, to build rapport, my 'in' was through responding to participants' material culture close at hand. Participants were introduced to the project and our ground rules for participation both practically and metaphorically through the experience of picnicking where we also discussed ground rules for participation (foregrounded in our motto 'don't yuck someone else's yum!').

Picnicking, for many, connotes conviviality, inclusiveness and delight. A material and embodied practice, it is a popular trope in representations of childhood. While the geographies of picnicking are still to be written, there are some excellent examples of edible cartographies out there. I invited participants to make edible cartographies ('Belfast buns'). This involved us getting elbow deep in icing, dough and mishmash recipes of the hyper local, supportive of edible avenues of inquiry that enabled participants to feel and munch their way through myriad layers and registers of feelings about their local area that can be difficult to pin down. We really enjoyed exploring what we could do with the stickiness using our hands and bodies. We loved blending colours to discover new ones, too, and agreed that marbling effects can be quite enchanting: 'now my Belfast is the universe!' cried one nine year old girl.

Button mapping

Eric was shocked to discover a button on the pavement – perhaps because of the uncanny likeness to a face. This easy-to-explain but powerful activity recognises ordinary objects like buttons as having the potential to cross-cut boundaries; to 'spark new thinking, connect people together and translate between them' (Steele, 2015, p.6). It was originally developed by Sue Moffat, director at the New Vic Theatre, alongside researchers involved with the Community Animation and Social Innovation Centre (CASIC) at Keele University.

I presented children in small groups a hoard of 'odds and ends' as we called it – not just buttons – lots of other bits and bobs migrated into our piles too – including things they collected or found. With the button mapping, participants were invited in small groups of 2–3 to do three things along with Eric:

- First sort the buttons (*this was open-ended – by colour? Decoration? Size? Button and non-button, etc.*).
- Next, organise and label the buttons, based on how they think their city is organised.

- Lastly, organise and label the buttons, based on how they and Eric thought Belfast should be organised in the future.

Buttons are linked to emotional geographies on both a metaphorical *and* physical level: they are familiar, non-threatening – to many – intimate even. They can be invested with lots of emotions. They can evoke memories associated with other people / places. They are open to interpretation. In contemplating and handling these little curiosities, in debating and interpreting and 'reading' each other's maps, participants enacted imaginative engagement with *their* Belfast, unsettling conventions of the real without losing sight of it, thus expanding and experimenting with boundaries of material everydayness in ways that disavowed predictability and tidiness (Pacini-Ketchabaw et al., 2017). Across all sites, it struck me that the literal transferability and open-endedness of these 'loose-part' materials (Nicholson, 1971) allowed participants to experiment and play with things and ideas – making *big* statements that they could always go back on, because they were not literally 'fixed' to place as it were. Indeed, I found participants across all sites to be exceptionally responsive to this cut. It worked particularly well in terms of practically introducing key mapping tropes like bird's-eye viewedness, juxtaposition, borders and relationality. Speaking in role as Eric in the final stage, the majority of participants found it quite easy to voice their thinking because of their shared affinity with the story and his character: he elicited their empathy, and respect.

Summary

We have enjoyed doing other pantry-inspired work, including 'pickling places' (using jam jars, water, glitter and *lots* of fimo clay) to gather, cluster and preserve special aspects of local areas that matter. The idea behind this cut was that participants, along with Eric, would explore and re-make places that mattered to them in miniature, with a view to 'preserving' them. Before getting our hands messy, however, we went walking *with* Eric in role as local tour guides to really get close to those aspects of the local areas that mattered. Across each of the creative cuts, participants' mapping and mapmaking have emerged as experimental and mutable modes of encounter, engagement and expression that *enact* moments of social, material, and personal imaginaries related to geographies of everyday matter/ing.

Conclusion: making the case for picture book as method/ology

In this chapter, I discussed Eric's expanded field of enquiry as extolling attentiveness (and responsiveness) to place detail. His is a narrative of difference, dynamism and change most pertinent to the theoretical and grounded approaches to research with / on childhood and youth as discussed by authors throughout this edited collection. *Eric* is, moreover, a manifesto for experimenting with and doing place-based research with children and young people differently. I close this chapter with an invitation. In the spirit of 'doing children's geographies' *differently* (Horton & Kraftl, 2005, 2006a; Van Blerk & Kesby, 2008), there is much scope I argue for creative and generative crossovers between children and youth geographies and children's literature. Indeed, children and adults alike are predisposed to stories and storytelling. Through both art and story, we can tell how things appear to be to us and explore how they could be different (Bradford et al., 2008; Somerville, 2013). Moreover, research is all about telling stories: stories that affirm the complexities of lives lived, never still, always on the move: through interviews, photographs, oral histories, etc. And yet, although popular in child education and literacy development in institutional contexts, the use of story book and picture book as method/ology is considerably under-explored in place-based research with children and young people to date, with the notable exception of Karen Murris' (2016) recent work. It is conspicuously absent within research on children and youth geographies especially.

Stories shared and passed on can impact on children and young people's differently nuanced ways of experiencing and experimenting with, playing and coping with their world. Picture books themselves are, like buttons, boundary spanning. Moreover, there are a range of exceptional picture books that can effectively broach concepts that researchers may find tricky to explain: gender relations / imbalances, sustainability, austerity, sectarianism; segregation and more, all of which impact on the lives of children and young people in many ways. Picture books permit a very special kind of connection-building where the written word need not take precedence. Rather, they enable and encourage particular attentiveness to visual languages as well as thing-play where props might be introduced, for example in role play and puppetry. In listening and looking *together* (a special form of collaborative inquiry) researcher and participant(s) can together shape space and build new directions in thought.

I have briefly outlined ways in which I use *Eric* in my fieldwork in Belfast as a lens through which participants explore, map and play with place. Through our workshops and activities, participants inscribe themselves in the 'thingness' and in what I call the 'little geographies' of their Belfast, responding to the interplay of the literal and the lyrical in tales that can enliven, enchant, politicise and *transform* place. Indeed, responsiveness to story as both medium and method can in turn set in motion different ways of making sense of place significance, exploring peculiarities, invoking new place wisdom, and uncovering those obscured or 'othered' bodies and things 'beneath' visible forms of landscape as it were (Tolia-Kelly, 2007). A storied approach to place-based research celebrates the world as full of entryways and doors, as Robert MacFarlane and Shaun Tan both recognise: a world that is always already becoming unfamiliar, decentred, connected and disconnected; a *smooth* space, open to the unexpected, the imaginative and lyrical, where new contacts and associations await. This is a very liberating thought. Let us respond then, and swing wide the doors to meet those different approaches and opportunities; to explore those lived experiences that await where story beckons.

References

Barad, K. (2003). Posthumanist performativity: Toward an understanding of how matter comes to matter. *Signs: Journal of Women in Culture and Society*, 28(3), pp.801–831.

Barad, K. (2011). Nature's queer performativity. *Qui Parle: Critical Humanities and Social Sciences*, 19, pp.121–158.

Benjamin, W. (2002 [1999]). *The Arcades Project*, trans. H. Eiland and K. McLaughlin. Cambridge and London: Belknap Press.

Bennett, J. (2010). *Vibrant Matter: A Politcal Ecology of Things*. Durham: Duke University Press.

Boyd, C. (2017). *Non-Representational Geographies of Therapeutic Art-Making*. Cham: Palgrave.

Bradford, C., Mallan, K., Stephens, J. and McCallum, R. (2008). *New World Orders in Contemporary Children's Literature: Utopian transformations*. Basingstoke: Palgrave Macmillan.

Bulsen, E. (2007). *Novels, Maps, Modernity: The Spatial Imagination*. New York: Routledge.

Clarke, H. and Witt, S. (2018). Findspots: Exploring children's everyday responses to real world geography. In: Geographical Association Annual Conference and Exhibition. University of Sheffield, 5–7 April.

Crossley, N. (1995). Merleau-Ponty, the elusive body and carnal sociology. *Body and Society*, 1, pp.43–63.

Deleuze, G., and Guattari, F. (1987). *A Thousand Plateaus: Capitalism and Schizophrenia*. (Original work Mille Plateaux, volume 2 of Capitalisme et schizophrénie published in 1980.) Translation and foreword: B. Massumi. Minneapolis: University of Minnesota Press.

Duhn, I. (2012). Places for Pedagogies, Pedagogies for Places. *Contemporary Issues in Early Childhood*, 13(2), pp.99–107.

Fox, N.-J. and Alldred, P. (2015). New materialist social inquiry: Designs, methods and the research-assemblage. *International Journal of Social Research Methodology*, 18(4), pp.399–414.

Gaiman, N. (2011). In conversation: Neil Gaiman talks to Shaun Tan. *The Guardian*. [online] Available at: www.theguardian.com/books/2011/dec/02/neil-gaiman-shaun-tan-interview [Accessed 25 Apr. 2018].

Groes, S. (2011). The spirit of discovery: A word of welcome from the psychogeographical travel agency by Sebastian Groes. [online] Available at: http://blog.museumoflondon.org.uk/the-spirit-of-discovery-a-word-of-welcome-from-the-psycho geographical-travel-agency-by-sebastian-groes/ [Accessed 25 Apr. 2018].

Hamm, C. (2015). Walking with place. *Canadian Children*, 40(2), pp.56–66.

Horton, J., and Kraftl, P. (2005). For more-than-usefulness: Six overlapping points about Children's Geographies. *Children's Geographies*, 3(2), pp.131–144.

Horton, J., and Kraftl, P. (2006a). What else? Some more ways of thinking and doing 'Children's Geographies'. *Children's Geographies*, 4(1), pp.69–98.

Horton, J., and Kraftl, P. (2006b). Not just growing up, but going on: Materials, spacings, bodies, situations. *Children's Geographies*, 4(3), pp.259–276.

Horton, J., and Kraftl, P. (2017). Rats, assorted shit and 'racist groundwater': Towards extra-sectional understandings of childhoods and social-material processes. *Environment and Planning D: Society and Space*. doi:10.1177/0263775817747278

Hultman, K. and Taguchi, H.L. (2010). Challenging anthropocentric analysis of visual data: A relational materialist methodological approach to educational research. *International Journal of Qualitative Studies in Education*, 23(5), pp.525–542.

Ingold, T. (2004). Culture on the ground: The world perceived through the feet. *Journal of Material Culture*, 9(3), pp.315–340.

Ingold, T. (2007). *Lines: A Brief History*. Routledge: London.

Ingold, T. (2008). Bindings against boundaries: Entanglement of life in an open world. *Environment and Planning*, 40, pp.1796–1810.

Ingold, T. (2011). *Perception of the Environment*. New York: Routledge.

Latham, A. and McCormack, D.-P. (2004). Moving cities: Rethinking the materialities of urban geographies. *Progress in Human Geography*, 28, pp.701–724.

Latour, B. (2004). *Reassembling the Social: An Introduction to Actor Network Theory*. Oxford: Oxford University Press.

Le Roy, F. (2017). Ragpickers and leftover performances. *Performance Research*, 22(8), pp.127–134.

Lorimer, H. (2005). Cultural geography: The busyness of being 'more-than-representational'. *Progress in Human Geography*, 29(1), pp.83–94.

Macfarlane, R. (2015). *Landmarks*. London: Penguin Books Ltd.

Malone, K. (2016). Theorizing a child-dog encounter in the slums of La Paz using post-humanistic approaches in order to disrupt universalisms in current 'child in nature' debates. *Children's Geographies*, 14(4), pp.390–407.

Malone, K. (2018). *Children in the Anthropocene: Rethinking Sustainability and Child Friendliness in Cities (Palgrave Studies on Children and Development)*. London: Palgrave Macmillan.

Manning, E. (2007). *Politics of Touch: Sense, Movement, Sovereignty*, Minneapolis: University of Minnesota Press.

May, E.T. (1989). Cold War – warm hearth: Politics and the family in postwar America. In: S. Fraser, G. Gerstle (eds.), *The Rise and Fall of the New Deal Order*. New Jersey: Princeton University Press, pp.153–185.

Murris, K. (2016). *The Posthuman Child: Educational Transformation through Philosophy with Picturebooks*. Oxon: Routledge.

Nicholson, S. (1971). The theory of loose parts: How not to cheat children. *Landscape Architecture*, 64, pp.30–34.

Nxumalo, F. (in press). Geotheorising mountain-child relations within anthropogenic inheritances. *Children's Geographies*, 15(5), pp.558–569.

Pacini-Ketchabaw, V., and Clark, V. (2014). Following watery relations in early childhood pedagogies. *Journal of Early Childhood Research*, 14(1), pp.98–111.

Pacini-Ketchabaw, V., Kind, S., and Kocher, L.M. (2017). *Encounters with Materials in early Childhood Education*. Abingdon: Routledge.

Pacini-Ketchabaw, V. and Taylor, A. (eds.) (2015). *Unsettling the Colonialist Places and Spaces of Early Childhood Education*. New York & London: Routledge.

Pyyry, N. (2017). Thinking with broken glass: Making pedagogical spaces of enchantment in the city. *Environmental Education Research*, 23(10), pp.1391–1401.

Rautio, P. (2013). Children who carry stones in their pockets: On autotelic material practices in everyday life. *Children's Geographies*, 11(4), pp.394–408.

Sacks, S. and Zumdick, W. (2013). *Atlas of the Poetic Continent: Pathways to Ecological Citizenship*. Forest Row, East Sussex: Temple Lodge Publishing.

Shirlow, P. (2006). Belfast: The 'post-conflict' city. *Space and Polity*, 10(2), pp.99–107.

Solso, R.L. (1996). *Cognition and the Visual Arts*. Massachusetts: MIT Press.

Somerville, M. (2013). *Water in a Dry Land: Place-Learning through Art and Story (Innovative Ethnographies)*. New York: Routledge.

Somerville, M. (2015). Emergent literacies in 'the land of do anything you want'. In *Children, Place and Sustainability*, S.M. Green (ed.). Palgrave Macmillan: London and New York.

Somerville, M. (2016). The post-human I: Encountering 'data' in new materialism. *International Journal of Qualitative Studies in Education*, 29(9), pp.1161–1172.

Springaay, S. and Zaliwska, Z. (2015). Diagrams and cuts: A materialist approach to research-creation. *Cultural Studies ☒ Critical Methodologies*, 15(2), pp.136–144.

Steele, J. (2015). *Boat Tree Game Button: How ordinary objects can spark extraordinary change. Cultural Animation and Community Organising Practice Report.* [online] Available at: www.keele.ac.uk/media/keeleuniversity/ri/risocsci/thelegacyof connectedcommunities/Boat_Tree_Game_Buttons_Practice_Report_single_page_print_updated%20(3)%20(1).pdf [Accessed 25 Apr. 2018].

Tan, S. (2009). *Tales from Outer Suburbia*. Dorking: Templar.

Tan, S. (2014). *Comments on Tales from Outer Suburbia*. [online] Available at: www.shauntan.net/about.html [Accessed 24 Apr. 2018].

Taylor, A. (2017). Beyond stewardship: Common world pedagogies for the Anthropocene. *Environmental Education Research*, 23(10), pp.1448–1461.

Thrift, N. (2008). *Non-Representational theory: Space, politics, affect*. Abingdon: Routledge.

Tolia-Kelly, D.P. (2007). Fear in paradise: The affective registers of the English Lake District landscape re-visited. *Senses and Society*, 2, pp.329–351.

Van Blerk, L. and Kesby, M. (eds.) (2008). *Doing Children's Geographies: Methodological Issues in Research with Young People*. Oxon: Routledge.

Waterton, E. and Dittmer, J. (2014). The museum as assemblage: Bringing forth affect at the Australian War Memorial. *Management and Curatorship*, 29(2), pp.122–139.

Woodyer, T. and Geoghegan, H. (2013). (Re)enchanting geography? The nature of being critical and the character of critique in human geography. *Progress in Human Geography*, 37 pp.195–214.

Yazbeck, S.-L., and Danis, I. (2015). Entangled frictions with place as assemblage. *Canadian Children*, 40(2), pp.22–31.

13 Transnational practices and children's local lives in times of economic crisis

Michael Boampong

Chapter summary

With weak economic growth indicators since the 2008 financial crisis and consequently widening social and economic inequalities regarding access to basic necessities and social services, one of the key areas that has not been sufficiently addressed in recent research is the transnational implications for households and the role of children. Drawing on data from multi-sited ethnographic research carried in Ghana and the United Kingdom, this chapter discusses transnationalism by focusing on the everyday life of families through the lens of young people's mobilities and transnational practices – the broad migration patterns that emerge including rural-urban and return mobilities. Movements and the everyday practices embedded within the transnational social field of the households reveal young people's agential competence in network formation, either by cultivating new ties or strengthening existing ones as part of how, individually or collectively, families negotiate the constraints of economic crisis. The analysis, which draws on Levitt and Glick Schiller's transnational framework on 'ways of being' and 'ways of belonging' in transnational social fields, also serves to make key contributions towards how we can further understand young people's distinctive role in transnational relationships and practices in times economic crisis.

Introduction

Transnational social fields, a theory utilised by scholars to better understand transitional migration, defines fields as a network of relationships through which people sustain ties, organise, exchange and transform ideas, practices, and resources (Faist 2000, p.189; Vertovec 2003; Schiller & Levitt 2004; Levitt & Jaworsky 2007; Faist et al. 2013). Vital functions within the transnational social field are the exchange of money and goods, cultural practices, values and people's movement across nations as part of everyday life. With that, most migrant families negotiate their survival amid political-economic welfare changes, immigration reforms, balancing production relations, and the everyday tasks necessary to run a household. The contributions of Olwig

(2002, p.216) in transnational migration studies has inspired theoretical contributions on the need to move beyond ethnic and homeland-diaspora relations to explain 'the practice of home, a domestic household unit'. Within this domestic unit, it has become important to explore children's agency (Parreñas 2005; White et al. 2011; Coe 2012).

In this chapter, I utilise fieldwork data on the everyday lives of Ghanaian transnational households, with relations living in Ghana and the United Kingdom (UK) to explain the role and meaning of the family. I will underscore how *doing family* in terms of the organisation and understanding of family is essential for child and parent survival of economic, social and political constraints. In this regard, certain children are incorporated in local, national and transnational social fields, especially in periods of economic change, including unemployment, rising cost of living and childcare, and lack of resources. Moreover, the concept of transnationalism, and to a lesser extent translocalism, are used, given that non-migrant children are embedded in both local and cross-border practices and social imaginaries. This discourse depends on Schiller and Levitt's (2004, pp.1010–1011) unique categorisation of transnational social networks and the consciousness of being embedded in transnational social fields as part of 'ways of being' (engagement in social relations and practices) and 'ways of belonging' (practices that signal or enact identity to a group, country or community).

A critique of dominant notions

The terms 'transnational family' and 'global householding' are used widely in transnational migration literature to conceptualise migrants' transnational behaviours, shifting the migration discourse beyond labour mobility and remittances into understanding the maintenance of family formations across borders. Families in which parents or children become migrants are known as *transnational families* (Bryceson & Vuorela 2002, p.3). They may live separately at different places and yet maintain a 'feeling of collective welfare and unity, namely familyhood, even across national borders' (Bryceson & Vuorela 2002, p.3). On the other hand, a global household is defined as comprising the people who migrate, people who are born into or otherwise incorporated into the immigrant household (e.g., through marriage or cohabitation) and people left behind in the origin country (Douglass 2013). Both terms connote how families adapt to changes due to mobility and migration. However, while transnational familyhood tends to emphasise the social and emotional capacity of the family, the global household stresses how the movement of people requires the household as a space of production and consumption to adapt to new ways of functioning materially. In this chapter, I will use *transnational household* for understanding the incorporation of 'other children' in the household, thus moving the focus beyond kin to include non-kin social relations as key 'players' (Mahler 1998, p.82) and relevant to social reproduction processes within Ghanaian transnational households.

It is important to note that traditional migration literature has been criticised for taking children's roles in migration processes for granted. First, children's influence in decision making has received limited attention. Despite this, evidence suggests children's preferences, constraints and aspirations may influence parental migration motives (Orellana et al. 2001; Parreñas 2005; Coe 2011b). Second, there is the tendency to assume migration is the undertaking of adult labourers (White et al. 2011). Thus, children are considered economic dependents and 'the luggage' of parental migrants (Orellana et al. 2001, p.578; Crivello 2015, p.39). In contrast, evidence by Orellana et al. (2001) has emphasised children's active roles in migration under various forms of mobilities and transnational practices. Moreover, to illustrate young people's independent migration, Hashim (2005), Kwankye et al. (2009), Hashim and Thorsen (2011) and Thorsen (2014), have all explored the situation of independent child migrants in Ghana and across borders in Africa. Their findings challenge the categorisation of unaccompanied child migrants as vulnerable, and consequent policies branding unaccompanied migrants as 'problematic' or 'out of place' (White et al. 2011, p.1163; Crivello 2015, p.39). Finally, transnational migration research suggests children *must* be kin members of a family to be engaged in transnational practices (Schiller & Levitt 2004). For instance, the focus on children of migrants or migrant children in receiving countries often tends to underestimate non-migrant and non-kin children and youth's subjective experiences, agency and resiliency, independent of adult involvement (cf. Orellana et al. 2001).

Studying family life and migration

Research context

Ghana

Historically, migration is a key survival strategy and a means of economic and social advancement for many Ghanaians. Ghana has been a principal emigration country in West Africa, where travelling abroad is 'one of society's cultural ideals' (Manuh 2005, p.7).

Internal migration is a critical feature of the Ghanaian household's livelihood strategy. People move for trading, missionary work and government official postings, and for securing cash crop farmland (Quartey 2009; Coe 2011a; Coe 2012). Internal migration, particularly rural-urban migration from northern to southern parts of Ghana, has been a significant migratory flow (Hashim 2005). Movements internally are linked to inequalities in spatial development as well as to poor colonial era and post-independence economic restructuring policies, including structural adjustment policies (SAP) (Awumbila 1997; Awumbila et al. 2014). Data from the Ghana Living Standards Survey (GLSS) suggests 48.6% of the population (24,658,823) migrated internally, with the majority of internal migrants living in Accra and Kumasi (Ghana Statistical

Service 2013a). Generally, females outnumber males in internal intra-regional migration (Ghana Statistical Service 2013a; Statistical Service (2013b). The surge in female mobility, especially among young unmarried women, is associated to men's increasing inability to meet the culturally demanded responsibility of being breadwinners. Thus, girls and young women move from rural locations to major cities to engage in either paid economic activities such as petty trading, *kayayei* [1] or unpaid domestic work in rich households.

International migration trends in the 2010 Ghana Population and Housing Census report (Ghana Statistical Service 2013a), indicated 250,623 Ghanaians live abroad, representing 1% of the country's population (24,658,823). Additionally, the 2010 census reported the dominant direction of Ghanaian emigration has been towards Europe – representing 37.7% of emigrants. Outside Africa, the bulk of Ghanaian emigrants live in major cities including London, Amsterdam, Hamburg and New York. These migration trends are concomitant with historical, colonial era, political and kinship ties, as well as advanced and cheap means of mobility (Anarfi et al. 2000; Manuh 2005). Generally, Ghanaians migrate for reasons including employment, education or training, marriage and family reunification, and political persecution (Anarfi et al. 2000; Coe 2013).

However, the literature also links high emigration in the 1970s and 1980s to the sharp economic decline and the introduction of SAP during the 1980s (Anarfi et al. 2003; Coe 2013). Consequently, SAP resulted partly in two outcomes: first, the introduction of austerity measures, including government staff redeployment and public expenditure reduction for social services, which incited people, especially the middle class, to migrate for better opportunities (Anarfi et al. 2003; Coe 2013). Second, men's economic deprivation in families motivated female mobility as an alternative livelihood strategy (Awumbila 1997).

Most emigrants originate from the Ashanti Region (27.6%) and Greater Accra (25.2%), mostly young adults in their 30s once they can afford travel costs (Ghana Statistical Service 2013a). The Ashanti Region and Greater Accra are major administrative regions with high youth unemployment rates (Quartey 2009) and established travel intermediaries (also known as 'connection men'),[2] who offer travel advice to facilitate visa acquisition (Coe 2013, p.98). The culture of migration has resulted in migration networks at home and abroad (Quartey 2009). Migration is not just a linear model; some migrants do return home for various reasons. Besides homesickness, emigrants return to Ghana for work-related reasons, marriage, or family reasons, or when migration objectives have been met (Coe 2013). Children may be sent home for foster care or eventual return to their parents' 'homeland' due to various political, sociocultural and economic constraints in the host country, which I will discuss with a focus on the UK (cf. Coe 2013).

The UK

Historically, the UK was the favourite destination of most Ghanaian emigrants, often for education and work, until the 1970s when there was a shift in immigration policy. Before the 1970s most Ghanaians moved through visa on arrival schemes; thus, most could travel to the UK without a visa. However, since the 1970s most countries, including the UK, decided to impose visa requirements for Ghanaian emigration. Most recently, the UK government passed a new family reunification law, requiring increment proof of financials (£18,600 per annum) to be demonstrated by British / settled people wishing to sponsor the immigration of a non-European (Gower and McGuinness 2015; Sumption & Vargas-Silva 2016). Therefore, movement out of Ghana to the UK has become more challenging for lower- and middle-income families.

The 2011 UK Census reported about 62,896 Ghanaian people living in London (about 0.8% of the total population of London) (Office for National Statistics 2014). Like other African migrants, some Ghanaian migrants take on jobs in the UK below their education or training experience (Coe 2013, p.117). Often this happens because of poor social capital, in addition to certain employer's unwillingness to recognise qualifications from Ghana. Still, migrant worker earnings remitted to Ghana often help in the acquisition of property and offers a middle-class standard of living for non-migrants or returnees (Coe 2013).

Having noted the social and economic positions of Ghanaians in the UK, it is also important to emphasise the political-economic conditions. Since the 2007/8 global economic crisis, fiscal austerity measures – mainly in the form of spending cuts and small tax increments – have had potential consequences on households' cost of living in Britain (Marie Hall 2016). Within the past ten years, government austerity policies and tax imposition include: a freeze on Child Benefit; constraining the working-age benefits to 1% per year; cutbacks in Local Housing Allowance and Housing Benefits and the setting of a Welfare Cap; removal of Tax Credits meant to compensate low-wage earning workers (Richardson 2010; Ridge 2013; HM Treasury 2010); rising cost of childcare; and increments in Value Added Tax from 17.5% to 20% (Ridge 2013). These expenditure and welfare cuts have not been proportional to pay raise or average wage growth (Clegg 2018), despite the fact that cost of living continues to increase (Posen 2012; Dowling and Harvie 2014).

Methodology and positionality

Fieldwork was undertaken in four south-east London boroughs considered highly populated with Ghanaian migrants: Southwark, Croydon, Lewisham and Lambeth (Office for National Statistics 2014). In south-east London many economic, social and cultural activities highlight the ways of living and belonging as Ghanaian immigrants, including self-operated businesses like Ghanaian food and remittance services, established Ghanaian Hometown Associations, and Ghanaian churches. As a temporary resident of south-east London myself, I was able to recruit respondents, share space and spend time with interviewees. In my daily mobilities to work, church and university, I met Ghanaians who had friendly conversations with me regarding life in London.

When I began data collection in London, I employed the snowball sampling approach and, by this, friends and acquaintances in London were useful introductions to potential interviewees.

Data elicitation methods were mainly informed by the benefits of ethnographic principles in research with children (James and Prout 1997). In brief, observation and prolonged interaction with research subjects were not only found to be useful for understanding everyday lives, but also in building trust and rapport, and breaking certain cultural barriers in adult-child relationships. In particular, Ghanaian culture teaches children to be subservient as a way of showing respect (Twum-Danso 2009; Boakye-Boateng 2010). Similarly, cultural notions, such as *a child must be seen and not heard* and if a child spoke he or she must be speaking *something right*, are characteristic of institutionalised power dynamics between children and adults. However, in spending more time with young people and their parents, many young people felt at ease speaking with me and sometimes engaged me in various activities such as play, schoolwork and housework. Additionally, it was emphasised that there are *no right / positive and wrong / negative answers* to avoid the tendency of *not speaking until I have the right answer.*

Simultaneous matched sample method (SMS) was utilised to link London-based respondents with people they maintained some form of transnational practices with living in the Ashanti Region (Mazzucato 2008). Hence, this research is marked by three key phases: the first phase was executed in London between March and June 2016 and consisted of participant and non-participant observations and semi-structured in-depth interviews. The second phase involved visiting Ghana between July and October 2016 to interview relatives of the London-based respondents to understand their everyday lives. Since returning to London, I have had some additional research interactions with my London respondents.

Participants in this research include 36 children (n=36, including 21 girls and 15 boys) aged between 5 and 19 years: 17 of these children live in London with their parents. Most of the respondents included adults age 30 or older, and their children who were between the ages of 7 and 30 years. Nineteen migrant parents were also interviewed in London. My interviewees in Ghana included children of migrants, returnee child-migrants and parents, and non-kin children living in transnational households located within towns in the Ashanti Region. In Ghana, I met 19 children referred as potential interviewees by their relatives in London. Additionally, I interviewed nine adults in Ghana who had caring responsibility for child interviewees; most caregivers were female. However, there was also a sense of intergenerational and intragenerational caring responsibility for the child from other members of the household, especially from the eldest sibling, uncles, aunts and their grandmothers.

Research conversations[3] were conducted in either Twi[4] or English depending on each participant's preference and language competence. All interviews in Ghana were held in Twi and the majority of parents in London interacted with me in Twi, only speaking in English at specific points when parents wanted to clarify or add to a child's response. Data in Twi was translated into English with care to avoid losing the implied and symbolic meanings of certain statements. Pseudonyms have been used to anonymise and protect the identity of interviewees.

As a Ghanaian, prior to my field work, I considered myself as an insider, given for instance my personal experience as a Ghanaian migrant, being Ashanti and having fluency in Twi. While these markers were advantageous, they also shifted with respect to people and place. In fact, there were some methodological challenges with being an 'insider'. First, my position as a Ghanaian seemed to make some people feel ashamed to tell me about their life struggles. Secondly, some, especially those who were undocumented, expressed signs of distrust in my position as a researcher given the growing sense of anti-immigrant policies. In many cases, trust had to be built by being introduced through friends as trustworthy, and through regular interaction and involvement in respondents' daily lives. Familial relationships evolved with some children and parents calling me 'uncle' – as a sign of respect – though in some cases it signified a sense of obligation. Children in poor families in Ghana, for instance, felt I was a wealthy migrant who could *save* their economic situation with material resources. For instance, Monica, an 8-year-old girl, and her mother asked me to send them money and a laptop upon my return to London. In other cases, power dynamics changed when children tried to teach me something (e.g. pulling a tuber crop, drawing water from a well, etc.). Laughing at me on such occasions, they suggested that I am not only an outsider, but also an incompetent adult. In essence, it became problematic to position myself strictly as an outsider or insider. Thus, with these experiences of being a Ghanaian, but viewed as a non-Ghanaian, or being viewed as a fictive adult, I feel I am situated in the *hybrid insider-outsider* status as discussed in the work of Carling et al. (2013, p.16).

Children and patterns of migration: two empirical cases

In order to more deeply explore transnational social fields, I will begin by describing two families from my research and their nature of organisation and activities. I will focus on the children's distinctive role as social actors in the process of forming 'new' networks and *doing family* as a way of navigating the effects of economic crisis.

Family 1: from rural-urban child migrant to becoming a transnational household member

The household consists of Geraldine (56-year-old single mother, living in London), Stella (Geraldine's 13-year-old daughter, living with Geraldine in London) and Geraldine's 'other child' Ama (17-year-old female, living in Kumasi).

Internal migration by young people, mainly from rural to urban communities is a common practice in Ghana. Ama moved from a village in Esaase-Bontefufuo to Kumasi, the capital city of Ghana's Ashanti Region. She described the village as a place where she had no access to good education, walked long hours to school, and after school had to rush to the farm to join her mother in weeding and to carry firewood to the house. She found farm work difficult and dangerous. Insufficient food required her to depend on fruits like pawpaw in the farm. On arrival in Kumasi, she lived with her mother's family friend as a 'house-help,' engaging in daily activities including sweeping, cleaning, fetching water and selling drinking water. Being the eldest daughter, she occasionally sent a local money transfer to her mother to support her mother and sibling's basic needs. In 2016 when I met her, she had gained admission into senior high school.

Geraldine met Ama in 2014 during a visit to Ghana where Ama was selling water. Geraldine had pitied Ama after listening to her village life experience. Ama, in a later conversation, also told me about her struggles with city life, including insufficient food and long working hours.

Like most migrants, Geraldine has built a house in Kumasi with no one living in it from day-to-day. However, when Ama told her about some of the challenges she faced in Kumasi, Geraldine decided to take responsibility for her – a process that is common in Ghana. Ama currently lives in Geraldine's unoccupied house. Her caretaking responsibilities of the house have meant 'there is life in the house';[5] she cleans and goes on errands for Geraldine, as Geraldine does not have a 'trusted' direct relation in Ghana to serve as caretaker. In Ghana, when an extended relation lives in the house of a transnational migrant, there is the fear over a claim of ownership or non-payment of rent by the occupant. Thus, instead of entrusting kin with the house, Geraldine trusts and relies on an adult-child network and a non-kin relation to maintain her '*borga fie*' (house built with migrant remittances or resources).

Ama refers to Geraldine as her *mother*, given that she is 'more caring than her biological mother, she takes care of me like her own child, keeps in touch every day, offers advice, asks me if I have eaten and how I am doing.' In the four years since she met Geraldine, Ama has become a member of their transnational household with future aspirations of joining them in London. Likewise, Geraldine sees Ama as her 'other' child – '*akwaada baako no*' as she constantly referred to her.[6] As the '*other* child', Ama proactively uses various tools, including language, as part of maintaining her position in the transnational household and therefore *doing family*. She sends biblical and motivational messages to Geraldine every morning. She changes her *WhatsApp* profile picture to that of Geraldine or Stella (Geraldine's daughter whom she has never met) regularly and occasionally, including on Mother's Day, includes messages such as 'Sweet Mother' and 'My Sweet Sister'.

Being part of the transnational household, Ama receives remittances from London for food and mobile phone credit, as well as school funds, which is essential given the rising cost of public education. In particular, in 2016 on my return to Ghana, Geraldine sent me with money, clothing and a power-bank for Ama, making it cheaper and more accessible for her to send remittances to Ama. It can be argued that, without Ama's social incorporation into the transnational family, these material resources would have been unobtainable.

Family II: sending the child back to Ghana to a non-kin social relation

I first met Florence (a 38-year-old woman) and her daughter, Afia (a 7-year-old girl) at the 2016 Kumasi-Tech Association summer barbecue in London. Afia came home after school and met me after I had finished her mother's interview. Her mother asked her to 'go and greet uncle,' and in doing so Afia remarked, 'oh you are my lost uncle.' Among Ghanaians, children recognise both kin and non-kin relations to indicate respect for an adult, however, being referred to as 'lost uncle' suggested Afia's expectation of my regular presence in her house. Additionally, this may serve to indicate children's interest in a researcher being part of their daily lives.

Afia's mother, Florence, sent Afia to Ghana as a young child. Afia was born in the UK but went to Ghana at the age of three and returned to London when she was six. Florence, as a single parent, working as a security guard with low income, felt it was near impossible to accomplish her economic goals, including full-time work and sending remittances to her father, while also caring for Afia (cf. Coe 2013, pp.115–116). With these constraints, Florence decided to arrange for a friend's mother to care for Afia in Ghana. She applied for a Ghanaian passport for Afia's travel to Ghana and to avoid any Ghana visa applications for future travels. This suggests that, amid time and financial constraints as well as high childcare costs, Florence responsibly and consciously chose a non-kin network to perform 'what it means to be a mother'; that is 'cooking, engaging with the child's school, discipline, reading bed-time stories, and essential child-rearing activities', all of which Florence felt she was unable to accomplish in London. While this emphasises expectations of a *good mother*, it also points to the constraints that migrants face in belonging to and being settled in London. This everyday focus is also manifested in how Afia viewed intimacy with her mother while growing up in Ghana. Her mixed emotions and therefore her emotional attachment emerged more strongly than the economic goals presented by her mother:

> I don't understand why my mother took me to Ghana and left me there at such a young age when I needed her most… I liked the toys she sent to my friends and me. She also called often on Skype, but sometimes after speaking with her, I doubted if she was my mother as she was not here [in Ghana] with me… I was taken to Ghana and I am back in London. I learned Twi language, and my grandmother[7] taught me how to speak Twi and respect people …but I miss her [my grandmother].

When children move to Ghana, they learn to speak their parent's local Ghanaian language and receive moral and cultural lessons that contribute to building cultural capital. This includes practices-based lessons such as saying '*Medaase*' (thank you), using the right hand in social interactions, and respect for adults. When children fail to demonstrate these, they are considered disrespectful, spoilt and non-Ghanaian by adults. For Afia to learn and understand Twi was fundamental to Florence in constructing Afia's identity beyond being British. Florence highlighted the importance of this shared cultural experience in everyday life and interaction with Afia: 'now since she came back from Ghana, we are able to communicate better in Twi. It makes it easier whenever I am with her; sometimes when you speak English even in the bus, everyone knows what you are saying but at least I am able to keep some secret with her through our local language and she is able to speak with other Ghanaians'.

Afia's caregiver, an elderly woman whom I met in Ghana, had a very pragmatic view of her family's relations to London and migrants:

> It is not easy to train *abrokyire nkwadaa* [overseas-born Ghanaian children] ... Florence and others tell me you cannot beat a child because your child can be taken from you and also they don't have much time to take care of the child. For here sometimes some people pamper *abrokyire nkwadaa* but I disciplined Afia when she did something wrong. I was sad when she left, she was crying when Florence came for her. It's sad Florence also hardly calls me on phone since she left for London.

Local activities and transnational practices in everyday life

Internal migration and, in particular, the phenomenon of rural-urban migration, represents one response to the day-to-day challenges of the lack of basic necessities – such as food – inequality, access to quality education and other services (Coe 2013, p.64). Child migrants to major cities like Kumasi may enter into petty trading activities, or be vulnerable to exploitation. As a result, they are sometimes beaten by city guards, or their items are confiscated to inform them they are 'out of place' and unneeded in the city (see, for example, Pearson, this volume, for a discussion of the stigmatisation of street-involved youth in Tanzania). In some ways, the experience of Ama contests notions about rural-urban migrants in Ghana. Ama, and the other child migrants I interviewed, suggested they made informed decisions drawing on social networks within the city. Thus, it is important to understand why children move and consider the networks of protection, including families, children may form as part of their mobilities.

The family experience of communication and togetherness underscore children's roles in transnational *ways of being* and *ways of belonging* from everyday life experience. The process of *doing family* by being physically together or being apart but connected via WhatsApp technology, and the imagination of *being* together through shared images and calls, also underscores the interlinking of distance and closeness, as in the case of Ama. There were times I observed her making calls in the *safety and comfort* of Geraldine's house in Ghana. Though miles apart, calls, text messages and the changing of profile pictures could be described as a way of being 'connected as a family', even though they are not blood family, but rather a 'new' family she has cultivated (Schiller and Levitt 2004, p.1011). The presence of memories, nostalgia and imagination with the images and communication were critical to this family formation, as well as showing care about others' wellbeing. The phone calls allow for the presence of Geraldine as a mother to her 'other child' at a distance and vice versa for Ama. Additionally, calls allow Geraldine to know the situation of her property which she only sees twice in a year when she visits Ghana.[8]

In contrast to Ama and Geraldine's experience with distance and closeness, Afia's caregiver lamented that she misses Afia, and that Florence no longer calls as frequently as she used to when Afia was in Ghana. This comment suggested familial bonds had developed through child movement. The child's role and agency in constructing and maintaining networks, especially in the age of cost-effective communication apps, became obvious only after the child returned to the UK. In relocating children to Ghana for child-rearing, children's contacts contribute to building connections in Ghana for both the parent and the children; children's presence in Ghana is a significant reason parents may return or keep in touch with relations or people in Ghana. For Afia, being taken care of by a foster parent created a personal connection to 'grandma' and, first and foremost, the ability to communicate in Twi. This can also orient children towards seeing Ghana as 'home'. Perhaps an issue outside the scope of this paper is whether children will really shift this latent transnationalism to active transnationalism along their life course.

Rather than representing children's mobility to Ghana as an act of abandonment by the parent, the experience of Florence, for instance, suggests a way of *reworking* motherhood and what care for a child should mean. The two family experiences in this chapter underscore the relational effect of changes in the political economy and how the transnational household's composition and connections play out. Additionally, as argued by Olwig (2007) and Coe (2013), places can no longer be seen as politically defined by borders, but rather cultural phenomena are critical to the meanings we attach to places within the transnational social field (cf. Schiller and Levitt 2004). With the practice of sending children to Ghana, the movement between Ghana-UK transnational social fields disrupts dominant categorisation of countries as 'host' or 'origin' places based on wealth and income indicators. Ama's active practice of *doing family* can be viewed as an orientation towards transnational imaginations of being together with Geraldine in London. Her use of online images suggests a *way of being* together and is a sense of a transnational way of belonging and being embedded in the transnational family or social field.

Concluding remarks

Key benefits of the transnational approach include the possibility of simultaneously capturing everyday life experiences of both moving and non-moving subjects and understanding how familial relationships between people and places are maintained in periods of economic change. However, as noted by Mahler (1998), the transnational lens may not reveal all transnational dimensions and actors. By using an ethnographic approach, the researcher is able to build trusted relationships with families that reveal consequences of political economics on issues such as access to basic needs and the rising cost of childcare. These issues help to inform transnational practices which impact on children's mobilities and relationships. Attention is brought to human agency, as young people are emplaced in new families within transnational households. Consequently, practices within transnational households challenge the dominant construction that to be a 'good mother' or 'good child' requires a nuclear family. In truth, transnational family networks serve to protect and care for children as families negotiate political economy constraints, and children eventually recognise non-kin relations as 'mother' and 'grandmother'. Further, for internal child migrants in Ghana, working and receiving remittances from overseas allows them to engage in translocal practices that highlight the globalised connection between families (Holt and Holloway 2006).

Notes

1 Young females who migrate to southern Ghana to work as head porters.
2 It is often a service procured by low- to middle-income people who have weak social networks at institutions that facilitate the acquisition of these documents.
3 While I explained to my respondents that this study was for my PhD, I often used the term *research conversation* in subsequent interactions with my respondents. The term conversation seemed appropriate given that interactions were often done while *on the move* with research subject(s). For instance, while walking to a shopping mall or while on a bus travelling together.
4 Being one of the major Akan languages spoken widely by most Ghanaians as a first and second language.
5 Among Akans it is believed that when a house is empty without occupants evil spirits may take over, or the house may collapse.
6 Literal translation of 'Akwadaa baako no'. This twi expression was used by Geraldine when she referred to Ama during our interactions to suggest a fictive kin or tie with Ama. It was a conscious effort of distinguishing between her blood children and non-kin child. That being said, Afia simply referred to Geraldine as her *mother* without using 'other mother' in her responses.
7 Afia's recognition of an older woman she lived with as 'grandmother' suggests the construction of a relationship between the child and her carer rather than a blood relation.
8 Of course, Ama and Geraldine's story raises important questions, though outside the scope of this chapter, about 'unpaid' child labour in the form of transnational housekeeping, given the growing significance of migrant property acquisition in Ghana to diversify income in order to survive rising costs of living in London.

References

Anarfi, J.K., Awusabo-Asare, K. & Nsowah-Nuamah, N. (2000) *Push and Pull Factors of International Migration: Country Report – Ghana.* Luxembourg, Publication Offices of the European Union.

Anarfi, J.K.*et al.* (2003) *Migration from and to Ghana: A Background Paper.* Brighton, University of Sussex: Development Research Centre on Migration, Globalisation and Poverty.

Awumbila, M. (1997) Gender and structural adjustment in Ghana: A case study in northeast Ghana. In: Awotona, A., & Teymur, N. (eds.), *Tradition, Location and Community: Place-Making and Development (Ethnoscapes).* Brookfield, Avebury, pp. 161–172.

Awumbila, M., Owusu, G. & Teye, J.K., (2014) *Can Rural-Urban Migration into Slums Reduce Poverty? Evidence from Ghana.* London, Migrating out of Poverty.

Boakye-Boateng, A., (2010) Changes in the concept of Childhood: Implications on children in Ghana. *The Journal of International Social Research*, 3(10), 104–115.

Bryceson, D.F. & Vuorela, U. (2002) *The Transnational Family: New European Frontiers and Global Networks.* [Online] Oxford, Berg. Available from: http://site.ebrary.com/id/10019857 [Accessed January 22, 2016].

Carling, J., Erdal, M. & Ezzati, R. (2013) Beyond the insider–outsider divide in migration research. *Migration Studies*, [Online] 2(1), 1–19. Available from http://migration.oxfordjournals.org/content/early/2013/10/15/migration.mnt022.abstract [Accessed June 20, 2015].

Coe, C. (2011a) How children feel about their parents' migration: A history of the reciprocity of care in Ghana. In: Coe, C., Reynolds, R.R., Boehm, D.A., Hess, J., Rae-Espinoza, H. (eds.), *Everyday Ruptures: Children, Youth and Migration in Global Perspective.* Nashville, Vanderbilt University Press, pp. 97–114.

Coe, C. (2011b) What is love? The materiality of care in Ghanaian transnational families. *International Migration*, [Online] 49(6), 7–24. Available from: doi:10.1111/j.1468-2435.2011.00704.x

Coe, C. (2012) Growing up and going abroad: How Ghanaian children imagine transnational migration. *Journal of Ethnic and Migration Studies*, 38(6), 913–931.

Coe, C. (2013) *The Scattered Family: Parenting, African Migrants, and Global Inequality.* Chicago, University of Chicago Press.

Crivello, G. (2015) 'There's no future here': The time and place of children's migration aspirations in Peru. *Geoforum*, 62, 38–46. Available from: http://linkinghub.elsevier.com/retrieve/pii/S0016718515000809.

Douglass, M. (2013) Global householding and social reproduction in migration research. *Ewha Journal of Social Sciences*, 29(2), 5–68.

Dowling, E. & Harvie, D. (2014) Harnessing the social: State, crisis and (big) society. *Sociology*, 48(5), pp.869–886.

Faist, T. (2000) *The Volume and Dynamics of International Migration and Transnational Social Spaces.* New York, Oxford University Press.

Faist, T., Fauser, M. & Reisenauer, E. (2013) *Transnational Migration.* Cambridge, Polity Press.

Ghana Statistical Service (2013a) *2010 Population and Housing Census: National Analytical Report.* [Online] Available from: www.statsghana.gov.gh/docfiles/publications/2010_PHC_National_Analytical_Report.pdf [Accessed June 20, 2015].

Ghana Statistical Service (2013b) *2010 Population and Housing Census Report: Children, Adolescents and Young People in Ghana*. [Online] Available from: www.statsghana.gov. gh/docfiles/publications/2010phc_children_adolescents_&young_people_in_Gh.pdf [Accessed June 20, 2015].

Gower, M., & McGuinness, T. (2015) *The Financial ('Minimum Income') Requirement for Partner Visas. Briefing Paper: SN06724*. London, UK House of Commons.

Hashim, I. (2005) *Research Report on Children's Independent Migration from North-eastern to Central Ghana*. [Online] Available from: www.migrationdrc.org/publica tions/research_reports/ImanReport.pdf [Accessed June 20, 2015].

Hashim, I. & Thorsen, D. (2011) *Child Migration in Africa (Africa Now)*. Uppsala, Zed Books Ltd.

HM Treasury (2010) Chancellor announces reforms to the welfare system. Press releases, GOV.UK. Available at: www.gov.uk/government/news/chancellor-announ ces-reforms-to-the-welfare-system [Accessed February 1, 2016].

Holt, L. and Holloway, S.L. (2006) Editorial: Theorising other childhoods in a globalised world. *Children's Geographies*, 4(2), 135–142.

James, A. & Prout, A. (1997) *Constructing and Reconstructing Childhood: Contemporary Issues in the Sociological Study of Childhood*. London, Falmer Press.

Kwankye, S.O., *et al.* (2009) *Independent North-South Child Migration in Ghana: The Decision Making Process*. [Online] Available from: www.migrationdrc.org/publica tions/working_papers/WP-T29.pdf. [Accessed June 20, 2015].

Levitt, P. & Jaworsky, B.N. (2007) Transnational migration studies: Past developments and future trends. *Annual Review of Sociology*, 33(1), 129–156.

Mahler, S. (1998) Theoretical and empirical contributions toward a research agenda for transnationalism. In: Smith, M.P. & Guarnizo, L. (eds.) *Transnationalism from Below*. New Brunswick, Transaction Publishers, pp. 64–100.

Manuh, T. (2005) *At Home in the World? International Migration and Development in Contemporary Ghana and West Africa*. Accra, Sub-Saharan Publishers.

Marie Hall, S. (2016) Everyday family experiences of the financial crisis: Getting by in the recent economic recession. *Journal of Economic Geography*, 16(2), pp.305–330.

Mazzucato, V. (2008) Simultaneity and networks in transnational migration: Lessons learned from a simultaneous matched sample methodology. In: DeWind, J., & Holdaway, J. (eds.) *Migration and Development Within and Across Borders: Research and Policy Perspectives on Internal and International Migration*. Geneva, International Organization for Migration, pp. 69–100.

Office for National Statistics (2014) *2011 Census: Age Structure, Local Authorities in England and Wales*. [Online] Available from: www.ons.gov.uk/ons/datasets-andta bles/index.html?pageSize=50&sortBy=none&sortDirection=none&newquery=Age +structure%2C+local+authorities+in+England+and+Wales&content-type=Refer ence+table&content-type=Dataset [Accessed June 20, 2015].

Olwig, K.F. (2002) A wedding in the family: Home making in global kin network. *Global Networks*, [Online] 2(3), 205–218. Available from: doi:10.1111/1471-0374.00037 [Accessed June 20, 2015].

Olwig, K.F. (2007) *Caribbean Journeys: An Ethnography of Migration and Home in Three Family Networks*. Durham, Duke University Press.

Orellana, M.F.*et al.* (2001) Transnational Childhoods: The Participation of Children in Processes of Family Migration. *Social Problems*, [Online] 48(4), 572–591. Available from: doi:10.1525/sp.2001.48.4.572 [Accessed June 20, 2015].

Parreñas, R.S. (2005) Children of global migration: Transnational families and gendered woes. *Contemporary Sociology: A Journal of Reviews*, [Online] 35(5), 480–482. Available from: doi:10.1177/009430610603500518 [Accessed June 20, 2015].

Posen, A. (2012) Why is their recovery better than ours? (Even though neither is good enough). Available at: www.bankofengland.co.uk/publications/Documents/speeches/2012/speech560.pdf[Accessed June 20, 2015].

Quartey, P. (2009) *Migration in Ghana: A Country Profile 2009*. Geneva, International Organization for Migration. [Online] Available from: https://publications.iom.int/system/files/pdf/ghana_profile_2009_0.pdf [Accessed June 20, 2015].

Richardson, D. (2010) Child and family policies in a time of economic crisis. *Children and Society*, 24(6), pp.495–508.

Ridge, T. (2013) "We are all in this together"? The hidden costs of poverty, recession and austerity policies on Britain's poorest children. *Children and Society*, 27(5), pp.406–417.

Schiller, N.G. & Levitt, P. (2004) Conceptualizing simultaneity: A transnational social field perspective on society. *International Migration Review*, 38(3), 1002–1039.

Sumption, M. & Vargas-Silva, C. (2016) *The Minimum Income Requirement for Non-EEA Family Members in the UK*. [Online] Migration Observatory Report, COMPAS. Available from: https://migrationobservatory.ox.ac.uk/wp-content/uploads/2016/04/Report-Minimum_Family_Income.pdf [Accessed June 20, 2015].

Thorsen, D. (2014) Jeans, Bicycles and Mobile Phones: Adolescent Migrants' Material Consumption in Burkina Faso. In: Veale, A., & Dona, G. (eds.) *Child and Youth Migration Mobility-in-Migration in an Era of Globalization*. Basingstoke, Palgrave Macmillan, pp. 67–90.

Twum-Danso, A. (2009) Situating participatory methodologies in context: The impact of culture on adult–child interactions in research and other projects. *Children's Geographies*, 7(4), 379–389.

Vertovec, S. (2003) Migration and Other Modes of Transnationalism: Towards Conceptual Cross-Fertilization. *International Migration Review*, [Online] 37(3), 641–665. Available from: www.jstor.org/stable/30037752. [Accessed June 20, 2015].

White, A.*et al.*, (2011) Children's roles in transnational migration. *Journal of Ethnic and Migration Studies*. [Online] 37(8), 1159–1170. Available from: www.tandfonline.com/doi/abs/10.1080/1369183X.2011.590635 [Accessed June 20, 2015].

14 Conclusion

Nadia von Benzon

This conclusion intends to draw out from the chapters in the volume ways to move forward in the framing of intersectionality in the experience of children and young people.

This collection, or rather the RGS-IBG conference session from which this volume derived, was birthed through the notion of 'other' as a way of framing the experiences of children whose everyday lives are shaped by difference within the socio-spatial context in which they are growing up. This framing of difference, as both relational and contextual, provides a lens for descaling children's geographies (Ansell, 2009) in that it inherently requires that the localised micro-geography of a child's life is considered within the broader social and structural context of the place and the culture in which the child is growing up. Difference, as understood through the chapters in this volume, is also temporal. Being politically active (Hall and Pottinger), non-attendance at school (von Benzon), engaging in sexual relationships (McRobert), fighting wars (Markowska-Manista), and working for money (Willman), amongst other examples, are only markers of difference of personhood if we factor in a temporal dimension. In other words, these activities would not be particularly unusual if undertaken by adults, but become unusual because they are part of the everyday lives of the children these chapters depict.

The lives of these children, marked by difference due to their age, or perhaps engaged in activities that are not broadly considered part of a global or local imaginary of what it is to 'do' childhood, tell us something specific about what it is to be a child, and ergo, about social personhood. These children cannot be 'becomings' in the manner that Nancy Worth uses to describe the experience of visually impaired youth (Worth, 2009, following, Grosz, 2004), as looking toward a future of independence as visually impaired adults. For many of the children in this volume, the identity that shapes their everyday lived experience has no futurity, at least in the long term, as their identity is grounded in the biopolitics of their current age (see also, Horton and Kraftl, 2006). The identity of adult child soldier or adult youth political activist do not exist, the person becomes an adult soldier or a political activist, subsumed into these adult categories of being that no longer denote difference. Therefore, a child soldier, for example, cannot be constructed as a human becoming, but is rather a complete and fully formed geopolitical being, as a result of the role that their youth plays in the demarcation of their specific identity.

Where the markers of difference are thus chronological, the category of difference, at least on the micro-scale of the individual, is truncated and specifically located in the time and space of youth. What this means for these young people is that the nexus of difference that underpins their experience of the world is not only embodied in the sense that it is located within the body, but is also reified of, or from, the body. The physiology and the psychology of the individual (Ahmed, 2008) is not so much embedded with difference, but is itself the difference, that plays out in these young peoples' lives. As young people grow out of children's bodies, and begin to transition to the habitation of adult bodies, some sorts of difference may enter liminality before 'being left behind'. The law makes spaces for liminality, and collective consciences recognise 'grey areas'[1] that acknowledge a transition from childhood to adulthood as a messy zone of becomings, unbecomings and rebecommings. The danger here, however, is that whilst the identity of childhood difference may ultimately be subsumed into an adult identity of normalised being, the impact of childhood difference – particularly in terms of the risk of impediment to social, emotional and psychological development – may have lifelong ramifications (for example, see Haer and Bohmelt, 2015; Teicher et al., 2016). On the other hand, where childhood difference has not been stigmatised or objectively dangerous, childhood difference may translate into adult competencies, or social and cultural capital that will continue to have an impact on an individual's life.

Whilst for some of the children in this volume, the difference inherent in their everyday experience of childhood will be – at least outwardly – outgrown, for others, childhood is the period in which they learn about, and test the boundaries of, what will be a lifetime of difference. Bihari children living with ethnic and linguistic difference in Bangladesh (Afroze), disabled young people (Wilkinson and Wilkinson), children growing up with interconnected British-Tanzanian families and residency experience (Boampong) will know alterity as a life-long aspect of their identity. Still, however, for some of these children the notion of becoming may not fit their lived experience of aging. If we look to the discourse of new materialism, now fully embedded in geographical conversations around social life and experience (for example in children's geographies: Arvidsen, 2018; Merewether, 2018), difference, or at least the society's response to difference, presents an interruption to theories of human and non-human interaction. At its most basic, difference interrupts materialist theory because it acts as a barrier – at least at an individual scale – to the human in the human-non-human interrelationships, preventing or shaping the interactions between individuals and both their micro-scale geographies and wider environmental and social forces. The biopolitics of structural and social agency suggest that age alone should bring increases in an individual's powers of choice and self-directed action, but stigmatised experiences of difference will also go on to inhibit these children's experiences of adult agency.

This focus on difference or alterity demonstrates that increasing autonomy is not universally correlated to maturity, an experience well-rehearsed in neighbouring social sciences such as disability studies and critical race studies (Bjornsdottir et al., 2014; Hondius, 2014). Goodley et al. (2015; 2016) develop this discourse through the idea of a dis/human or dishuman child. The authors posit that not only does disability contradict a universalised understanding of what it is to be human, but has the potential to contribute to a reconceptualisation of this very phenomenon. Particularly, the authors critique the notions of autonomy and independence as features of humanhood. This argument logically redirects the focus of interrogation of 'agency' away from the individual, to understand agency as a characteristic of complex assemblages of humans and non-humans, or perhaps as the product of dynamic networks of interrelationships in which an individual is situated at any spatial or temporal moment. This recognition of the inutility of the notion of individualised agency reflects the critical discourse addressing neoliberalism and feminist social critique. To do otherwise, Holloway et al. (2018) argue, is to overlook the power of the social and political structures in which children experience their everyday lives.

Wolpert's (1980) 'Dignity of risk' calls for reflection on risk-taking, one aspect of autonomy, as a human right. Wolpert's paper argues that in contemporary society serious risk is mitigated with the intention of curtailment through, amongst other things, the removal of liberty of those with intellectual impairment. This action is taken, Wolpert argues, against those unable to maintain a 'cloak of competency' (Wolpert, 1980: 400). This cloak of competency, he states, is the privilege of those deemed 'normal', a category he notes that significantly grew in scope in the second half of the twentieth century as social attitudes to morality and personhood changed. As civil rights work and social activism continue to extend the scope of what is considered normal in contemporary (Western) society, then recognition of alterity, it follows, will be applied to a shrinking category of persons.

This broadening of normalcy reflects victory for social justice and, where advances reflect real social change, offers a revolution in the lived experience of individuals no longer stigmatised or marginalised as a result of perceived difference (Goffman, 1973). However, the apparent liberalisation of Western social values that underpins this broadening of the social construction of normalcy and difference does not spell universal liberation. For those not encompassed within the broadening category of 'normal', life might be experienced increasingly on the margins. As the mainstream increases in volume and spatial distribution, those considered 'other' are at risk of loss of allies, resources and structural protection and support. Moreover, social liberalisation is spatially bound, and the same broadening of acceptability experienced in Western countries has not been recognised in many parts of the world. Indeed, the very phenomena of Western social liberalisation may result in increased repression for some, as particular communities, political parties or States seek to push back against perceived threat to their own embedded cultural values. Given that difference can only be defined contextually, it is therefore incumbent upon geographers to position difference as socially and spatially relational, in order that identity-related experiences of difference, and particularly of associated stigmatisation and marginalisation, not be overlooked because the lived experience would not be one of alterity in another context.

Kraftl (2015: 219), in his discussion of alterity as it relates to the experience of children engaged in alternative education, argues for the use of the term 'alter-childhoods' as a tool for 'open ... and productive questioning', rather than as a static categorisation that critiques particular forms of normalcy. For Kraftl, alterity is not simply a being, but a doing, an activity that can be performed knowingly in order to reshape and reform childhoods in response to, or in spite of, a mainstream social construction of what it is to be a child. Meanwhile alterity is not described by Kraftl as a fixed identity, but rather as 'complex and in process' (2015: 234). The complexity of the definition of what might constitute alterity in childhoods mirrors a messiness around normalcy, in itself building to a broader concern of the utility of the notion of childhood and youth as a framework for understanding identity. If categorisation and therefore definition of what it is to be a child or a young person is slippery, and there is an apparent absence of heterogeneity within the messy, fragile, brittle and flaky identity grouping of youth, what can possibly be the benefits of discourse, let alone empirical research, that seeks to explore the lived experiences of young people?

And yet, the chapters of this book demonstrate that despite the differences in young people's lived experience that contrast with the normative social goings-on within their context, and contrast with the experiences of other young people within this volume, there exist similarities that conceptually bind young people as a distinct grouping from adults. In other words, the similarities inherent in the biopolitics of youth may in some contexts transcend other differences within this population. Mulvenna's chapter, exploring the work of novelist Shaun Tan, demonstrates the potential universalism of the experience of childhood as a period of unknowing and of familiarisation with the world around us. The naivety of Tan's protagonist, Eric, and their ability to reform consciousness through innocent exploration of the locality in which they found themselves, places an onus on others to recognise the subversive and innovative potential of the child. Here it is not so much the act of finding, sorting and displaying that holds agentic power, but the way in which the child's actions might be interpreted and acted upon by the adult viewer that contains potentiality. Rather than romanticise the potential agency of the child, this interpretation serves to underscore the extent to which children's apparent agency – their ability to work (Willman), their participation in politics (Hall and Pottinger; Willman), their engagement in supplementary or alternative educations (Wilson and Warren; von Benzon) – is only possible through adult permission to access space and resources. Machiavellian leadership theory suggests that the perception of personal agency is important to facilitate control and prevent rebellion. Thus offering young people a sense of control over the course of their lives, and allowing a sense of choice and self-directed actions may serve to support the regime in which the young person is situated, through limiting the likelihood of youth rebellion (see for example, Willman).

Over-emphasis on the power of childhood agency is also problematic in that it threatens to diminish or overlook the lived experience of people for whom the biopolitics of youth do not transcend other identity markers. A variety of literature, referred to in the introduction of this text, suggests that for young people of minority racial and ethnic groups and for young people with an LGBTQ+ identity, identities other than age may dominate in particular social or spatial settings (e.g. Rodó-de-Zárate, 2015). This idea of identity as fluid and contextual underpins the notion of intersectionality in which individual's perception of their own identity, and their own experience of difference, reflects the social and environmental context in which they are situated in any given moment. For the Bihari children in Afroze's chapter, for example, their identity as child may shape and constrain their lived experience within the Bihari enclave. However, when attending school outside this spatial area, their identity as child is normalised within the environment, and thus subsumed by their identity as Bihari which acts as a marker of difference, limiting their opportunities for socialisation amongst their non-Bihari peers.

To conclude, this volume has sought to reflect on alterity as a means of exploring, and contributing to, the developing critique within children's geography of the notion of agency, and to tentatively suggest a way forward. This path forward seeks to follow feminists and critical social theorists in denouncing the notion of agency as a capacity situated within an individual, and rather reframing agency as socially mediated and contingent upon the social value placed on identity and personhood. Personal choice and self-directed action can only be experienced in social and structural environments in which an individual's identity is recognised as capable. A child who 'leans in' may succeed in an environment that is in some way conducive to, or at least permissive of, this child's efforts. Indeed, a child who leans harder within this context may do better. However, no amount of personal effort or attempted action will increase a child's agency if the environmental context is unwavering in its hostility to the personhood of that child. Thus agency, understood as the ability to choose and to take direct action, needs to be framed as the product of complex interpersonal and socio-spatial relationships that amplify or silence a child's personal desires and needs.

Note

1 For example, in English and Welsh law sexual activity with a minor is treated differently depending on whether they are under or over 12 years old.

References

Ahmed, S., (2008). Open forum imaginary prohibitions: Some preliminary remarks on the founding gestures of the 'New Materialism'. *European Journal of Women's Studies*, 15(1), 23–39.

Ansell, N. (2009). Childhood and the politics of scale: descaling children's geographies? *Progress in Human Geography*, 33(2), 190–209.

Arvidsen, J. (2018). Growing dens. On re-grounding the child-nature relationship through a new materialist approach to children's dens. *Children's Geographies*, 16 (3), 279–291.

Bjornsdottir, K., Stefansdottir, V., and Stefansdottir, A., (2014). 'It's my life': Autonomy and people with intellectual disabilities. *Journal of Intellectual Disabilities*, 19(1), 5–21.

Goffman, E. (1973). *The Presentation of Self in Everyday Life*. Woodstock, NY: Overlook Press.

Goodley, D. and Runswick-Cole, K. (2015). Thinking about schooling through dis/ability. In: Corcoran, T., White, J., Whitburn, B. (eds.). *Disability Studies: Innovations and controversies (interrogating educational change)*. Rotterdam: SensePublishers, 241–253.

Goodley, D.Runswick-Cole, K. and Liddiard, K. (2016). The dishuman child. *Discourse: Studies in the Cultural Politics of Education*, 37(5), 770–784.

Grosz, E. (2004). *The Nick of Time: Politics, evolution, and the untimely*. Durham, NC: Duke University Press.

Haer, R., and Bohmelt, T. (2015). Child soldiers as time bombs? Adolescents' participation in rebel groups and the recurrence of armed conflict. *European Journal of International Relations*, 22(2), 408–436.

Holloway, S. L., Holt, L., and Mills, S. (2018). Questions of agency: Capacity, subjectivity, spatiality and temporality. *Progress in Human Geography*. doi:10.1177/0309132518757654

Hondius, D. (2014). *Blackness in Western Europe: Racial patterns of paternalism and exclusion*. Abingdon: Routledge.

Horton, J., and Kraftl, P. (2006). Not just growing up, but going on: Materials, spacings, bodies, situations. *Children's Geographies*, 4(3), 259–276.

Kraftl, P. (2015). Alter-childhoods: Biopolitics and childhoods in alternative education spaces. *Annals of the Association of American Geographers*, 105(1), 219–237.

Merewether, J. (2018). New materialisms and children's outdoor environments: Murmurative diffractions. *Children's Geographies*, 17(1), 105–117.

Rodó-de-Zárate, M. (2015). Young lesbians negotiating public space: An intersectional approach through places. *Children's Geographies*, 13(4), 413–434.

Teicher, M., Samson, J., Anderson, C., and Ohashi, K. (2016). The effects of childhood maltreatment on brain structure, function and connectivity. *Nature Reviews Neuroscience*, 17, 652–666.

Wolpert, J. (1980). The dignity of risk. *Transactions of the Institute of British Geographers*, 5(4), 391–401.

Worth, N. (2009). Understanding youth transition as 'becoming': Identity, time and futurity. *Geoforum*, 40(6), 1050–1060.

Appendix

Discussion and engagement resourceThe following questions have been devised to facilitate the use of the book as an educational resource. The questions might be used to spark discussions in seminars, tutorials or reading groups and are intended to encourage the reader to think reflexively about the content. Some of the questions might be considered appropriate for stimulating assessed coursework, essays or formative written work from students. The questions differ in the depth and level of analysis and reflection required. This resource has been developed with the needs of upper level undergraduate students and postgraduate students in mind, but may be largely suitable for other groups with the necessary support or adaption.

There are questions grouped here relating to each of the twelve substantive chapters of the book. The questions at the end of this resource are intended to be cross-cutting, first through each part of the book, and finally through the book as a whole.

Part I: Stigma

Chapter 2: Childhood disability and clothing: (Un)dressing debates

1   What sort of 'normalising' techniques, using clothing, are discussed by the authors as being used by a disabled child?
2   What challenges might there be in finding appropriate clothing for a child with an impairment?
3   How might inappropriate clothing stop a child with an impairment from fully engaging in everyday life?
4   How might a social view of disability guide designers in producing more appropriate clothing for a disabled child?

Chapter 3: 'They should have stayed': Blaming street children and disruption of the intergenerational contract

1   How are the children's street-based relationships discussed in this chapter different from the children's former relationships with their families and home communities?

2   In what way is van Blerk and Ansell's (2007) idea of 'intergenerational contracts' used by Pearson in this chapter?

3   What reasons does Pearson provide as to why street-living children experience stigmatisation in Tanzania?

4   Discuss the importance of relationships of reciprocity for street-living children.

Chapter 4: Subverting neighbourhood normalcy and the impacts on child wellbeing in Malta

1   In what ways, discussed throughout this chapter, are the health and well-being of people contingent to place?

2   Providing examples from the chapter, in what ways are children and adolescents affected by the prevalent gendered norms and processes transmitted in the neighbourhood?

3   In the 'traditional walled' neighbourhood discussed in this chapter, how do Roman Catholic religious norms impact on families?

4   How does the upholding of norms in the 'traditional walled' neighbourhood contrast to the 'modern' neighbourhood?

Chapter 5: 'Bad children': International stigmatisation of children trained to kill during war and armed conflict

1   Discuss the motto which begins the chapter that 'no child is born a murderer'.

2   Should child soldiers be considered victims?

3   How might the stigmatisation of child soldiers be harmful to the child soldiers themselves?

4   What is the role of the media in communicating the role and actions of child soldiers?

Part II: Work, Education and Activism

Chapter 6: Other(ed) childhoods: Supplementary schools and the politics of learning

1   How do supplementary schools respond to forms of structural discrimination?

2   How is supplementary education positioned in relation to mainstream education and other forms of alternative education?

3   In what ways, discussed in this chapter, has supplementary education been 'othered'?

4   Should use of supplementary schools be considered a result of, or reaction to, structural oppression?

Chapter 7: Not an 'other' childhood: Child labour laws, working children and childhood in Bolivia

1   How do the experiences of childhood for children in Bolivia differ from the normative view of childhood experienced in the global north?

2 Discuss the idea that children involved in child labour are 'helpless victims'.

3 What might a Bolivian child understand by the term 'responsible childhood'?

4 Bourdillon (2006) summarises two different approaches to childhood. One is that childhood is a time to be free from work and employment, and instead dedicated to learning, play and fun. Using material from Willman's chapter, critique this idea.

Chapter 8: Unschooling and the simultaneous development and mitigation of 'otherness' amongst home schooling families

1 In what way could a child who is unschooled be considered 'other' within a neoliberal context?

2 What techniques, discussed by von Benzon, do families employ to ensure that this 'otherness' is a positive experience for their unschooled children?

3 In what ways might the childhood geographies of unschooled children be distinct from their mainstream education peers?

4 In what way is the agency of children central to unschooling?

Chapter 9: Being seen, being heard: Engaging and valuing young people as political actors and activists

1 What challenges do young people face to be taken seriously as political actors and activists?

2 In what ways do the young people discussed throughout the chapter make sure they are 'seen and heard' as political actors?

3 How do the narratives presented in this chapter counter the idea of declining interest and disaffection between young people and politics?

4 In what ways does Team Future's demand 'instead of talking at us/for us, work with us' relate to notions of agency?

Part III: Out of Place

Chapter 10: Young survivors of sexual abuse as 'children out of place'

1 In what ways does McRobert position young survivors of sexual abuse as 'children out of place'?

2 How is the discussion of children's agency complicated in the context of sexuality?

3 To what extent does the character Una, in the discussed film *UNA*, have agency?

4 Does the category of a sexually abused child disrupt the idea of childhood itself?

Chapter 11: Realising childhood in an Urdu-speaking Bihari community in Bangladesh

1 To what extent is violence embedded in 'ordinary everyday lives' (Parkes, 2015, p. 2; see Chapter 11 References section)?
2 Discuss the idea that the relationships between children and adults are not fixed, drawing on the concept of 'child-adult relationships as relational processes'.
3 How do gendered power dynamics in the camp shape understandings of power and authority among children and adults?
4 What are the children's experiences of violence presented in this chapter?

Chapter 12: Discovering difference in outer suburbia: Mapping, intra-activity and alternative directedness in Shaun Tan's Eric

1 In what way can the character Eric in Shaun Tan's story *Eric* be considered 'other'?
2 How is difference explored in Eric's negotiation with place, space and things within an unfamiliar suburban environment?
3 In what way does *Eric* invite readers to identify, explore and critique attitudes towards strangeness, both in the story and real-life contexts?
4 How might *Eric* be used as a methodological approach to 'do' children's geographies differently?

Chapter 13: Transnational practices and children's local lives in times of economic crisis

1 How do practices within transnational households challenge the dominant construction that to be a 'good mother' or 'good child' requires a nuclear family?
2 Discuss the way in which the concept of transnationalism is used throughout this chapter.
3 How might internal migration be considered a critical feature of the Ghanaian household's livelihood strategy?
4 In what ways do the two family experiences in the chapter underscore the relational effect of changes in the political economy?

Questions relating to Part I: Stigma

Drawing on examples from across the four chapters in Part I:

1 Can stigmatised children ever diminish or eradicate stigma through their own actions, or does this always involve intervention from others?
2 In what ways can stigma impact on both the social and economic capital of children and families?

3   In what situations might the stigmatising impacts of difference present more challenges than the differences themselves, and when might this pose particular difficulties for those intending to improve the lived experience of a stigmatised child?

4   How does stigmatisation and marginalisation impact on a young person's use of space?

Questions relating to Part II: Work, education and activism

Drawing on examples from across the four chapters in Part II:

1   To what extent are the chapters in Part II reflecting failures of the State to meet children's and families' economic and social needs?

2   Does working and/or political engagement threaten or promote children's right to education?

3   Is it more important to protect young people's right to education and leisure, or promote young people's right to economic and political engagement?

4   How might these chapters call into question regulatory frameworks that restrict young people's engagement in political and economic life on the basis of date of birth?

Questions relating to Part III: Out of place

Drawing on examples from across the four chapters in Part III:

1   The chapters in Part III invite the reader to consider socio-spatial relationships from a child's perspective; what might be the challenges in doing this for academics or policy makers?

2   To what extent should the experience of violence or engagement in criminalised activity be considered a normal part of childhood?

3   Is 'goodness' in childhood synonymous with 'innocence'?

4   Should 'competent' children and young people have the right to make decisions about their own lives, regardless of parental opinion?

Broader questions addressing the whole book

1   The notion of inter-generational responsibility frequently reappears in the volume, with children positioned as capable of, and morally bound to, contribute to the household through their labour. With reference to some of the chapters in this book, explore whether the notion of inter-generational responsibility is compatible with the protection of the rights of the child?

2   Drawing on examples from different chapters, can children's agency be of consequence beyond the household in which they live?

3   Providing specific examples from this book, in what ways are the lives of these othered children underpinned by social and economic capital?

4   In what ways does the notion of stigmatisation differ in application across the chapters in the book?

We hope that these questions are useful in engaging groups in discussion or writing on some of the themes introduced throughout the book. The authors would be delighted to hear from anyone who has found the book useful in teaching or peer discussion.

References

Bourdillon, M., et al. (2006) 'Children and work: A review of current literature and debates', *Development and Change*, 37(6), pp.1201–1226.

Parkes, J., (2015) Introduction. In: J. Parkes, ed. Gender Violence in Poverty Contexts: The Educational Challenge. London: Routledge, 3–10.

Van Blerk, L., & Ansell, N. (2007) Alternative care giving in the context of Aids in southern Africa: complex strategies for care. Journal of International Development: *The Journal of the Development Studies Association*, 19(7), 865–884.

Index

Entries in *italics* denote figures.